TUGS OF WAR

HEAVY RECOVERY VEHICLES, TANK TRANSPORTERS,
and ARTILLERY TRACTORS
of THE BRITISH ARMY, 1945-1965

by
PAT WARE

•

WAREHOUSE
PUBLICATIONS

•

Published in Great Britain by Warehouse Publications.
© Copyright 1995 by Warehouse Publications.
First printed 1995.
ISBN 0-9525563-1-6.

All rights reserved. Apart from any fair dealing for the purpose of private study, research, or critical review, as permitted by the relevant parts of the 'Copyright and patents act 1988', no part of this publication may be reproduced or transmitted in any form or by any means, electronic, chemical, optical or mechanical, including photocopy, recording, or any information storage and retrieval system, without prior permission in writing from the publisher. All enquiries should be addressed to the Publisher.

Designed and produced by Warehouse Publications.
Page layout by Lizzie Ware.
Illustrated by Baxter & Brand, Radcliffe-on-Trent.
Text produced using 'Wordstar' V6.0; page layouts prepared using Aldus-Adobe 'PageMaker' V5.0.
Set in 10.5/11.5pt ITC New Baskerville.
Linotronic typesetting output by Tradespools, Frome, Somerset.
Printed in England by Staples Printers Rochester Limited.

Published by Warehouse Publications
5 Rathbone Square, Tanfield Road, Croydon CR0 1BT.
Telephone: 0181-681 3031.
Fax: 0181-686 2362.

TUGS OF WAR

Heavy recovery vehicles, tank transporters, and artillery tractors of the British Army, 1945-1965

CONTENTS

	FOREWORD	1
	INTRODUCTION	3
	Heavy tractors	5
1	**AEC**	9
1.1	Matador	11
1.2	Militant	19
2	**ALBION**	31
2.1	CX Series	33
3	**DIAMOND T**	43
3.1	Model 980/981	45
4	**LEYLAND**	55
4.1	FV1000 series	57
4.2	Martian FV1100 series	65
4.3	FV1200 series	93
5	**SCAMMELL**	103
5.1	Pioneer	105
5.2	Constructor	117
5.3	Explorer	129
6	**THORNYCROFT**	139
6.1	Antar	141
7	**TRAILERS AND SEMI-TRAILERS**	155
7.1	Full and semi-trailers	157
	INDEX	167

TUGS OF WAR

FOREWORD

I have to start by confessing that I can't resist big trucks, and that this book is an unashamed attempt to exploit that weakness.

Readers of my previous publication, 'In National Service' will probably recall my view that post-war British military vehicles are a much neglected field. So, once again I have confined myself to those vehicles in use with the British Army during the two decades following the end of World War 2, concentrating on tank recovery and transporter vehicles, artillery tractors, and recovery vehicles, rated at 10 tons or more.

It is a shame for admirers of large British military vehicles that the biggest of the 'big trucks' from WW2 is the American-built M26 Pacific 'Dragon Wagon'. With its 17.86 litre Hall Scott 440 petrol engine, and chain drive to massive Knuckey rear bogies, the 'Dragon Wagon' is probably the ultimate military heavyweight. You will find that I did manage to work it into the text somewhere, but since it really falls outside the parameters I set myself in planning the book, I have had to search for adequate substitutes among the British vehicles.

However... students of the 'big truck' will not be disappointed!

I have unearthed a lot of new photographs and material on the 60 ton FV1000, and 30 ton FV1200 vehicles. These huge wheeled tractors were planned by FVRDE to cope with what they saw as the ever-increasing rise in the size and weight of tanks. At more than four metres wide, and almost as high as a double-decker bus, the FV1000 particularly must have been an awe-inspiring sight on the tiny roads of post-war Britain.

Of the other post-war vehicles, I have also included the AEC 'Militant' in its three 'marks', the FVRDE-designed FV1100 Leyland 'Martian', the basically-commercial Scammell 'Constructor' and 'Explorer' tractors... and of course, the always-impressive Thornycroft 'Mighty Antar'.

While generally omitting those WW2-period vehicles which remained in service only because they survived the war, I make no apologies for also including four trucks which most certainly date from the war years.

The Diamond T was an obvious candidate since it remained in service until the 1970's and was up-engined in the 'fifties with the Rolls-Royce C6 diesel. I couldn't leave out the Scammell 'Pioneer' which, despite originally dating from the 'twenties, also remained in service well into the post-war years. And, well... the big Albion CX22S, CX24S and FT15N tractors just seemed to round the set off nicely and, in its way, one of these was almost reincarnated as post-war design. I have also included the experimental CX33 since this gets no coverage at all in other works on the subject.

And finally, with supplies being made both during and after WW2, the trusty AEC 'Matador' seems to have a foot in both camps.

I hope that 'big truck' fans will enjoy this work, and as always I would welcome notification of the inevitable errors or omissions... but don't let them kid you, size does matter!

Pat Ware
Croydon, 1995

MEMORANDA

Form of presentation
As with previous books, it was my intention to produce a reference work which could be accessed for specific information. I did not intend that the book necessarily be read from cover to cover like a novel. I have presented the text in what I believe is the best form to reflect this.

Information
I would like to thank all those who patiently answered my questions and helped me with my research. The list includes, in alphabetical order:

Brian S Baxter; REME Museum
Peter Gaine
Graham Holmes; Tank Museum
Laurence Spring; Surrey Record Office
Roslyn Thistlewood; British Commercial Vehicle Museum Trust Archives

Nomenclature
It seems there is no such thing as a definitive nomenclature, and for each vehicle, I have given what appears to be the 'lowest common denominator'.

Documentation
Each chapter concludes with a list of military technical publications and in some cases, suggested further reading. These lists are intended to be representative rather than comprehensive.

FVRDE
Up until the mid-1950's, the Ministry of Supply maintained separate research and development bodies with executive responsibility for the design and specification, and testing of so-called 'fighting vehicles'.

In 1954, the two executives were merged to form the Fighting Vehicles Research and Development Establishment (FVRDE). Although not necessarily historically accurate, for the sake of convenience, I have tended to use this description throughout the text.

It is worth pointing out that FVRDE and the Military Experimental Engineering Establishment (MEXE) merged in 1972 to form the Military Vehicles Engineering Establishment (MVEE), now renamed MoD Chertsey.

Photographs
I have tried wherever possible to use photographs which have not been seen before and I am grateful to those who allowed me access to their archives. With few exceptions, the black-and-white photographs are contemporary to the development and service of the vehicles. The colour photos were taken at various times over the last 15 years, and I would like to thank all who allowed me to photograph their vehicles.

The following codes are used to indicate the copyright holders of the photographs:

BCVM	The archives of the British Commercial Vehicle Museum Trust; Chorley, Lancashire
CVRTC	Commercial Vehicle and Road Transport Club
IWM	Imperial War Museum; London
KP	Ken Porter
MM	Mike Maslin
MP	MotorPhoto
NMM	National Motor Museum; Beaulieu, Hampshire
PG	Peter Gaine
REME	REME Museum; Arborfield, Berkshire
TMB	Tank Museum; Bovington, Dorset

Thanks also to the various manufacturing concerns for the use of their archive material.

INTRODUCTION
HEAVY TRACTORS

INTRODUCTION: HEAVY TRACTORS

INTRODUCTION
HEAVY TRACTORS

This book describes the heavy and medium/heavy tractors used by the British Army for moving tanks, field artillery, and engineers' plant, and for the heavy recovery and breakdown role, from the end of WW2 to the mid 1960's. The vehicles used in these two decades can be classified into three distinct groups, each with its own characteristics and philosophy.

Immediately after WW2 was over, the services were obliged to make do with vehicles which dated from the war years, and sometimes from before. Despite their often-poor condition, many of these vehicles remained in service throughout the period in question, surviving the misconceived attempts at producing purpose-built 30 and 60 ton 'combat', or CT, tractors as implementation of the post-war vehicle programme began.

These purpose-built post-war 'CT' vehicles form the second of the three generations. Designed by the Fighting Vehicles Research and Development Establishment (FVRDE) during the 'forties and 'fifties, with some input from the motor industry, the CT vehicles were a brave attempt to produce a range of purpose-built, 'pure' military vehicles embodying a high degree of component standardisation. With the benefit of hindsight, it's hardly surprising that the scheme foundered on grounds of excessive cost and unreliability, both of which can probably be attributed to lack of production volume.

Finally, as memories of the hard lessons of the war years began to fade, military thinking changed and a new generation of heavy vehicles, largely based on commercial designs, was successfully introduced into service. In the end it was these vehicles, rather than the FVRDE CT designs, which were to finally replace the WW2 trucks, many of which have only been retired in the last decade.

In order to put these vehicles into perspective, it is necessary to understand the situation which existed at the end of WW2.

THE PROBLEMS

Less than 20 years passed between the end of the Great War and the beginning of Hitler's military expansion into 'the former German protectorates', and it was to take six years of all-out global warfare, and the sacrifice of millions of lives to defeat Hitler.

Now, if only we would pay attention, history has much to teach us but since we are not generally keen students, all we seem to be able to learn is that history tends to repeat

INTRODUCTION: HEAVY TRACTORS

itself! And in that endless cycle of repetition, even before WW2 was over, Britain and America had every reason to believe that they would be at war in Germany once again, perhaps in as little as five years, this time with their former allies, the USSR.

But six years of all-out, world-wide warfare had demonstrated a few hard facts to the Allies about the logistics of mechanised warfare - lessons which had not been apparent at the end of WW1. Both Britain and the USA were keen that the lessons they had learned would be put into practice before the next big conflict... whenever that occurred.

Vehicle types and makes
The multiplicity of vehicle types and makes in use during the war years had proved to be a logistics nightmare.

The solution seemed obvious: reduce the number of types and makes involved, and ensure maximum standardisation between them. Since commercial vehicle designs were not generally appropriate, this forced the War Office to embark on an ambitious design programme with all the drawbacks that this entailed.

Weight of AFV's
Tanks and trucks had got bigger and heavier, and the transporter and recovery equipment had proved inadequate.

The answer? Make the transporters and recovery vehicles bigger and heavier. But of course this brought with it a whole new set of problems, not least of which was the lack of suitable motive power units, but also for example, problems relating to bridge capacities, manufacturing capabilities, and the availability of appropriate technologies and engineering skills.

Wading
The problems of getting vehicles ashore during the D-day landings had highlighted the inadequacies of appliqué waterproofing and it seemed that the only answer to this was to build-in deep-water wading capabilities at the factory.

The result of this, of course, was an enormous degree of, often redundant, technical complexity combined with difficulties of access for maintenance.

Electrical screening
Radio interference had also proved to be a problem, with those vehicles suppressed against interference having to be specially identified. Not all vehicles carried radios but all of them emitted radiation on radio frequencies from the ignition system and from motors and generators, and thus all of them created interference.

The Diamond T survived well into the 'seventies (REME)

Dizzy T's were sometimes double-headed to handle extreme loads (IWM)

The Martian was a military design later offered commercially (BCVM)

INTRODUCTION: HEAVY TRACTORS

The Explorer was offered to military and commercial customers (REME)

The Constructor was a commercial design modified for the army (BCVM)

... as was the hugely successful Antar (MP)

It seemed that this could be cured by ensuring every vehicle was radio screened but, of course, once again the downside was increased complexity.

Petrol vs diesel
Fuel supply had also proved extraordinarily difficult. Some vehicles required petrol, others only ran on diesel and getting the right fuel to the right place, at the right time, required valuable resources.

Since small, efficient diesel engines did not exist in the 1940's, the answer to this seemed to be to standardise on petrol engines... but, large petrol engines are inordinately thirsty, so yet again, one problem was simply replaced with another. And anyway, there seemed little point in standardising on petrol once engines were accepted into service which required 80 octane fuel, as opposed to the standard 70 octane.

Domestic production
And last, but by no means least, by effectively maintaining a blockade of the Atlantic for most of the War, the Germans had demonstrated beyond a shadow of doubt the foolishness of relying on imported goods and materiel.

There was only one answer to this particular problem: make the stuff in Britain.

THE VEHICLES THAT SURVIVED

With the exception of the American-designed Diamond T 980/981 tractors, those heavy vehicles that survived WW2 and remained in service through the early post-war years were essentially pre-war designs. In its military form, the Scammell 'Pioneer' dated from 1937, and the AEC 'Matador' from 1939; and although the Albion had been introduced in 1943, it was based on essentially pre-war technology.

There was actually nothing wrong with any of these vehicles as far as they went, and of course even by the end of the war, the designs were still less than 10 years old. But perhaps the trouble was that they did not go far enough for the way mechanised warfare seemed to have developed.

But, what they were able to offer was a rugged simplicity, combined with relative ease of maintenance... and in the headlong rush for sophistication which characterised the post-war vehicles, these much-needed virtues were forgotten.

THE ANSWER... OR WAS IT?

Before WW2 was over, the Chiefs of Staff had begun to draw up a plan for the post-war range of 'B' vehicles. Everything was to be standardised: weight classifications, engines, components, electrical systems, bodies, wading abilities. There were to be 30 ton and 60 ton tractors to cope with the increasingly heavy tanks. And everything was to run on petrol.

It seemed that this would solve all the problems.

The FVRDE design teams set to work with enthusiasm. Some idea of the levels at which military spending were running can be gained from the fact that the 1955 census of military vehicles showed that 52,500 new prime movers had been purchased since 1950. However, despite the rate at which new vehicles were being delivered, it wasn't long before disillusion began to set in. It became rapidly apparent that there was a number of problems.

The first problem was money, or lack of it. These vehicles were proving far more costly than had originally been envisaged.

Secondly, reliability. The FVRDE design teams lacked the experience and resources of the major manufacturers and some of their seemingly-innovative design solutions were not so easy to build.

Thirdly, time. All of this was taking far longer than anticipated and the 'red' threat from the East seemed to be looming ever closer.

And finally, customer satisfaction was also not running at high levels... the user arms were not at all happy with those vehicles that had been delivered.

So, it was obvious that it was necessary to go back to the drawing board, both literally and metaphorically. Those CT vehicles which had been delivered were downgraded to GS (general service) and the CT category was effectively abandoned. Several of the weight categories were dropped; others were retained but the proposed purpose-designed CT vehicles were replaced by modified commercial designs.

And, it was here that the whole purpose-designed vehicle scheme foundered, for the (GS) modified commercial designs were perfectly satisfactory. In many cases they were better than the proposed CT vehicles... and they were certainly cheaper.

CT vs GS

Finally we come to the third philosophy, the beginnings of which are well-represented by vehicles such as the Scammell 'Constructor', the Thornycroft 'Antar' and the AEC 'Militant'.

These were basically commercial designs modified to provide the type of enhanced performance demanded by the military, but with all of the advantages of proven design, and relatively-large volume production.

These modified commercial designs proved satisfactory and, although the philosophy has been refined, for example there is currently a trend for so-called 'multi-fuel' engines, it remains more-or-less unchanged to this day.

It is interesting to speculate whether the whole cycle would repeat itself were there to be another major conflict... but then history is supposed to repeat itself isn't it?

AEC
CHAPTER 1

CHAPTER 1: AEC

The Associated Equipment Company, better known by its AEC acronym, was created from the repair and maintenance workshops of the London General Omnibus Company (LGOC), and the Vanguard Company who had come together in 1907, forming what was eventually to become London Transport.

AEC was officially formed as a separate company in 1912 with a factory site at Walthamstow, and was established with the intention of building bus chassis for the LGOC, and its successors, as well as for the many hundreds of bus operators across the length and breadth of the land. It was not until the Great War broke out that the company diversified into commercial vehicle chassis, supplying more than 10,000 examples of the AEC 'Y Type' 3 ton truck to the War Office.

Whilst bus chassis production continued to grow, demand for trucks also increased after the armistice, even though the pace of progress was such that few designs remained viable for more than a few years. However, progress brought sales and, like many companies, AEC continued to go from strength-to-strength. In 1926 the company formed a two-year association with Daimler as the Associated Daimler Company (ADC). This was not to last, and in 1927, on its own again, AEC moved to a new 63 acre site at Southall where it was to remain until 1979.

A milestone was passed in 1930, when AEC introduced their first 'oiler', or diesel engine, developed in conjunction with the Ricardo Company. Rated at 8.1 litres, this was soon replaced by an 8.8 litre unit, along with smaller 5.3 and 6.6 litre versions. In 1937, a 7.7 litre unit was introduced, going on to become the company's standard for many years. With the famous 'Mammoth Major', AEC was also the first company to build an eight-wheeled truck with a rigid chassis.

At the outbreak of WW2, AEC turned the Southall factory over to war production, manufacturing 4x4 and 6x6 'Matadors' for use as cargo trucks, tankers, artillery tractors and mobile cranes, as well as in a variety of light armoured roles. The O853 4x4 'Matador' was rated as one of the best medium artillery tractors available to the Allies.

War production ceased in 1945, and although many 'Matadors' remained in service well into the 'sixties and even the 'seventies, it was not until 1952 that AEC received their first post-war military orders when contracts were placed for a total of 3000 'Militant' Mk 1 vehicles, in both 6x4 and 6x6 chassis configurations. More orders followed in 1953, when 'Matador' production was restarted, carrying on until 1959, by which time a further 1800 had been constructed.

In 1948, AEC absorbed Maudslay and Crossley, and although all of the chassis were to continue to be badged as AEC, the company changed its name to the Associated Commercial Vehicles Group (ACV). In 1961, the old-established Thornycroft company was also absorbed into the Group (see Chapter 6) and in August 1962, ACV merged with Leyland. Although the specialised military vehicles continued more-or-less unchanged, it was not long before the AEC commercial vehicles began to lose their identity with the adoption of the Leyland-designed 'Ergomatic' cab.

In 1966, the 'Militant' Mk 1 was replaced by the Mk 3, available only in 6x6 layout, and intended for use in tanker, cargo vehicle and recovery tractor roles. With their cabs based on the recently-discontinued Mk V commercial vehicle, it could be argued that these 'Militants' were the last of the true AEC's since, by the time they went into production, the commercial AEC's had already adopted the Leyland cab and were virtually indistinguishable from their Leyland stablemates. Production of the 'Militant' was continued into the 1970's but by this time the Leyland empire was in serious trouble.

Both the Army and the RAF continued to order AEC chassis during the difficult Leyland years and 'Mercurys', 'Mammoth Majors', and 'Mandators' could be seen in use as tankers, gas bottle carriers, aircraft de-icers, and missile transporters.

In 1977 the AEC name was dropped from commercial chassis, and in 1979 the Southall factory closed its gates for the last time, with all production being transferred to the former Leyland plants.

Of course this was not the end of the story, for in 1979, the incoming Conservative government refused to continue the cash handouts and the mighty Leyland edifice began to totter towards complete collapse. When the DAF takeover was signed in 1987, any hopes that AEC might emerge from the wreckage disappeared overnight, as did so many of the well-established names of British commercial vehicle building.

CHAPTER 1.1

AEC MATADOR
853, O853

'The British Army is the most satisfied customer AEC ever had... the 'Matador' is regarded as the best tractor in the medium class in either of the opposing armies.' So wrote the Commander-in-Chief of the Middle East forces in a message sent from Cairo to the AEC Company at Southall during the North African campaign of 1942 to 1943.

The model O853 4x4 'Matador' had its origins in a 4 ton four-wheel drive truck produced by Hardy Motors and the Four Wheel Drive Motor Company in the early 'thirties. Hardy Motors was subsequently absorbed into AEC where the work continued, and prototype 'Matador' chassis were first produced in 1937. The first military 'Matador' chassis was supplied to the British Army in 1938, intended for use as a medium artillery tractor. Normally powered by AEC's own 7.7 litre oil engine, the vehicle soon gained a reputation for enjoying hard work and abuse, and it wasn't long before the 3.7in heavy anti-aircraft guns which the 'Matador' was originally designed to tow, were also joined by the 5.5in howitzer and 7.5in field pieces.

The rugged and versatile 'Matador' was also adapted to other roles, and the chassis was used for cargo, armoured command, mine-laying, and gun portee variants. The RAF also used the 'Matador', employing many examples as a heavy tractor for drawbar trailers and specifying a 6x6 modification which was supplied as a fuel tanker, as well as being found in the armoured command post and mobile crane roles.

Production of the WW2 'Matador' finished at the end of 1945 - for students of 'Matador' minutiae, AEC tells us that it was actually at 2.45pm on 5 November in 1945 - after a total of nearly 10,500 vehicles had been produced. Many remained in service until well into the 'sixties and 'seventies, but the vehicle's willingness for hard work and appetite for downright abuse, made surplus 'Matadors' a popular choice with commercial operators, particularly as a breakdown/recovery vehicle, or showman's or forestry tractor.

The War Office must also have retained its high opinion of the 'Matador' into the 'fifties, for between 1953 and 1959, when deliveries of the new FVRDE-designed vehicles were not going according to plan, and despite the introduction of the more-modern 'Militant', a further batch of 1800 'Matador' chassis was purchased. These vehicles differed only in detail from their WW2 predecessors, and once again were bodied as artillery tractors, often being used to tow the 3.7in anti-aircraft gun, together with its generator and radar trailers.

In late 1958, AEC produced a brochure describing an updated Mk 2 4x4 'Matador' with what appeared to be the redesigned cab used on the 'Militant' Mks 2 and 3, and the AEC AV470 7.68 litre diesel engine. To all intents and purposes, this truck seemed to be virtually a 4x4 version of the normally six-wheeled 'Militant', and although it was apparently offered to military customers both in the UK and overseas, there were no examples purchased by the British Army.

There was also a range of Mk 2 and Mk 3 4x2 commercial 'Matadors', designated models O347 and 3471; these were quite different to the military models and were never in military service.

DEVELOPMENT

Before the Great War, engineer Charles Cleaver had been responsible for the design of the articulating rear bogie employed on the AEC 6x6 War Department chassis, and which, incidentally, remained in use more-or-less unchanged until the demise of AEC in 1977. After the war, Cleaver co-founded the Four Wheel Drive Motor Company, where, with Hardy Motors, he was involved in the design of a 4 ton 4x4 chassis which was eventually to become the AEC 'Matador' 4x4.

When AEC purchased the Four Wheel Drive Motor Company, Cleaver once again found himself employed by his old masters. He went on to complete the design work for the model 853 petrol-engined 'Matador', which, with its steel-roofed commercial-style cab, was basically a civilian design, and the O853 model which was fitted with an AEC oil engine.

In 1938, following a period of cross-country trials, which also included the products of other manufacturers, the Ministry of Supply placed an order with AEC for 200 'Matador' gun tractors with a load-carrying capacity of 3500kg and a maximum towing weight of 6500kg.

Aside from the very early examples which were fitted with an AEC petrol engine (and the 'commercial' style cab), all of the WW2 military 'Matadors' were fitted with the AEC A173 six-cylinder oil engine, producing 95bhp from a capacity of 7.58 litres. Vehicles supplied under the post-war contract employed the A187 engine, offering an additional 10bhp from the same capacity.

The rather 'perpendicular' cab, with its distinctive curved, canvas-covered roof, was designed to accommodate two, while the composite wood-and-steel rear body provided seating for a gun crew of nine or ten, with racks and stowage for their personal kit and the ammunition. Most

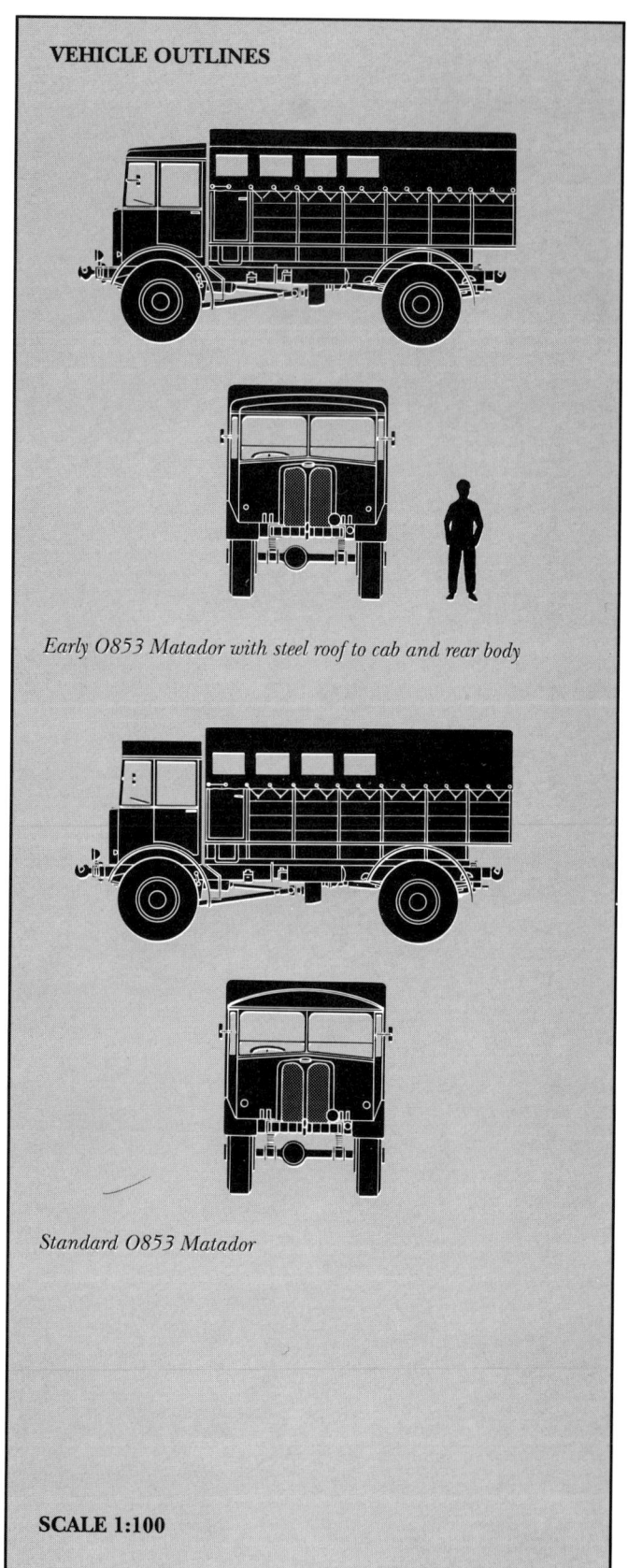

VEHICLE OUTLINES

Early O853 Matador with steel roof to cab and rear body

Standard O853 Matador

SCALE 1:100

CHAPTER 1.1: AEC MATADOR

SPECIFICATION

Dimensions and weight

	Engine: A173	A187
Dimensions (mm):		
Length	6325	6337
Width overall	2426	2388
Height:		
to top of cab	2946	2946
maximum height	3099	3073
Wheelbase	3848	3848
Track:		
front	1908	1908
rear	1797	1797
Ground clearance	330	330
Weight (kg):		
Laden	11,048	11,024
Unladen	7204	7532
Bridge classification	12	12

Performance

	A173	A187
Speed (km/h):		
maximum, on road	58	60
average, off road	20	20
Fuel consumption (litre/km)	-	0.26
Maximum range (km)	576	579
Turning circle (m)	18.3	16.8
Maximum gradient	-	44.5%
Approach angle	50°	50°
Departure angle*	42°/34°	42°/34°
Fording unprepared (mm)	762	762
Capacity (kg):		
Maximum load	3500	3500
towed load	6500	6500

* Figures quoted are for War Office/Admiralty, and Air Ministry vehicles respectively.

examples were also fitted with a 7 ton mechanically-driven winch mounted between the chassis rails and beneath the rear body.

The 'Matador' was the first 4x4 truck to be produced in Britain in quantity, and in all, some 10,411 'Matadors' were supplied over a 20 year period: 8611 during the war years, and a further 1800 of the post-war type. Although most of the 4x4 'Matador' chassis were supplied for use as artillery tractors, there were also cargo vehicles and a range of other types. Some 400 vehicles designated 'tractor, heavy, 4x4' were supplied to the RAF intended as tow vehicles for drawbar trailers; this may well have been the role for which the 'towing' version of the later Scammell 'Explorer' was intended.

There was also a 6x6 variant, based on the front end of the 'Matador' chassis, and the rear bogie of the 'Marshal'. In production form this version was only supplied to the RAF, but the chassis was also used to produce a prototype low-silhouette heavy artillery tractor. Ten examples of the prototype were produced, but there was no series production and the work was abandoned at the end of the war. However, the 6x6 chassis also formed the basis of the post-war 'Militant' Mk 1.

An interesting experimental, three-quarter tracked variant of the 4x4 'Matador' chassis was also produced during the war years, intended for towing the 17 lb and 6 lb field guns. The rear axle was removed and replaced by the complete track and suspension system from a 'Valentine' tank. Although similar vehicles were favoured by the Germans, the experiment was apparently not a success and was not continued.

It must have been obvious by 1953 that the 'Matador', which was essentially a pre-war design, was well past its best. However, when difficulties were encountered with deliveries of the new CT-rated Leyland 'Martian' (see Chapter 4.2), the Ministry of Supply fell back on the tried and tested 'Matador' and approached AEC with a proposal for restarting production to remedy what had become a shortage of suitable artillery tractors. Between 1953 and 1959, AEC produced a further 1800 examples of the O853 'Matador', with only detail differences when compared to the originals.

In May 1958, when the 'Martians' were frozen in depots awaiting clutch, suspension and auxiliary gearbox repairs, the Director Royal Artillery (DRA) proposed that they be replaced in service by the 'Matadors', of which there were apparently 'sufficient free stocks'. This was obviously not simply seen as a short-term stop-gap measure, for in October of that year, a memo was sent from the War Office to BAOR suggesting that the 'Martians' might remain frozen until 1960, and if this were the case, stating that it might be necessary to move 'Matadors' to Germany.

The 'Matador' was an extremely tough and reliable truck, remaining in service until well into the 'seventies. Some idea of just how tough it was, can be gained from the fact that, during the desert campaigns, at least one example was known to have been converted in the field to act as a breakdown/recovery vehicle by the addition of a Hyster fixed jib taken from a US recovery vehicle. This vehicle was perfectly capable of recovering disabled AFV's. The 'Matador' was also sometimes pressed into service as an emergency tank transporter, towing the 45 ton Rogers trailer intended for use with the Diamond T (see Chapter 3.1).

NOMENCLATURE and VARIATIONS

Tractor, 4x4, medium artillery, AEC Matador O853
Four-wheel drive artillery tractor with two-man cab, and composite wood-and-steel rear body designed for a crew of nine or ten, together with 3500kg of stores and ammunition. Early versions used a petrol engine, but most were equipped with a direct-injection diesel engine, either the A173 or A187 unit, each of 7.58 litres; four-speed gearbox and two-speed transfer case. Some examples fitted with 7 ton mechanical winch.

The 6x6 model was designated AEC 'Matador' O854.

DESCRIPTION

Engine

Although the very early 'Matadors' (model 853) were fitted with a petrol engine, during the greater part of the production run, there were two different diesel engines, or 'oil' engines as they were called at the time, fitted into the 'Matador' chassis, both of AEC manufacture.

The WW2 contract vehicles used the six-cylinder A173 engine, while vehicles supplied under the post-war contracts were fitted with the A187. In practice, the differences between the engines were small, but while the A173 was used in a number of different AEC chassis produced immediately before and after the war, the A187 was only used in the model O853 (and O854) 'Matador'.

Both engines were water-cooled direct-injection, four-stroke six-cylinder diesels, producing 95 and 105bhp respectively from 7.58 litres, at a governed engine speed of 1800rpm.

The block was a one-piece casting, fitted with renewable dry cylinder liners. Twin cylinder heads, each covering three cylinders, were cast from fine-grained iron. The crankshaft was carried in seven main bearings. Overhead

Matador prototype with flat-bed body, 1937/38 (TMB)

Early artillery tractor with steel-roofed cab and rear body (TMB)

Rear view of same vehicle - note rear wings compared to prototype (TMB)

CHAPTER 1.1: AEC MATADOR

Line-up of Matador trucks with drivers - 'somewhere in England' (AEC)

Well-restored Matador in WW2 guise - note front towing pin, 1995

Restored cargo truck shows off its 'Mickey Mouse' ears camouflage, 1995

valve gear was driven by pushrods from a low-mounted camshaft. The fuel-injector pump and governor were supplied by Simms, while the actual injector nozzles were manufactured by CAV; an Amal lift-type pump was used to supply fuel to the injection equipment.

Engine data

	A173	A187
Capacity (cc)*	7580	7580
Bore and stroke (mm)	105x146	105x146
Compression ratio	16:1	16:1
Gross power output (bhp)	95	95/105
Maximum torque (lbs/ft)	310	330
Firing order	142635	142635

* There is a slight mystery here; several reliable sources, including AEC's own sales literature, while still quoting the bore and stroke as 105x146mm, claimed that the swept volume was 7.71 litres rather than 7.58 litres. Similar discrepancies exist with other AEC engines, and this may well be due to the way AEC themselves calculated engine capacities.

Cooling system

The engines were both water-cooled in the conventional manner using a centrifugal pump-assisted, atmospheric thermo-siphon system in conjunction with a 'Still' tube radiator.

A 560mm diameter six-bladed fan, driven by means of belts and a slipping clutch, was used to draw air through the radiator.

Transmission

Power was transmitted through a 400mm single dry-plate clutch, to a four-speed and reverse gearbox, unit-constructed with the engine. First and reverse were engaged by a sliding gear, whilst the other speeds were of the constant-mesh helical type engaged by sliding dog clutches.

There was a separately-mounted, two-speed auxiliary gearbox connected to the main gearbox by a short, solid propeller shaft. The auxiliary gearbox contained the transfer gear assembly, with the front-wheel drive interlocked with the low ratios, and also provided drive for the winch. The transfer box to axle ratios were 1:1 and 2.3:1.

A Westinghouse or Clayton Dewandre compressor was mounted on the main gearbox to provide compressed air for the braking system.

From the auxiliary gearbox, an open propeller shaft ran the length of the vehicle to drive the rear axle; a second, shorter shaft ran from this to the second rear axle on 6x6 chassis. A third propeller shaft ran from the transfer case to the front axle. AEC's own couplings were used on the shaft connecting the auxiliary gearbox to the main gearbox and to the winch, with the more usual Hardy Spicer type used on the main driving propeller shaft joints.

Suspension and axles
The suspension at both front and rear was by means of long semi-elliptical multi-leaf springs attached to the chassis by shackles and pins, and to the axles by standard 'U' bolts.

Final drive to the wheels was transmitted through a double-reduction spiral-bevel/double helical-gear type differential within each axle casing, and then, on the rear wheels, through fully-floating axle shafts. Bendix 'Tracta' type constant-velocity joints were used in the front axle to provide driving and steering actions.

The axle ratios were 7.9:1, although the later WW2 contract vehicles used a worm axle at the rear with a 7.75:1 ratio, with changes made to the transfer gear ratios.

Steering gear
The chassis was of the forward-control type, with the steering box mounted ahead of the front axle. Steering was controlled through a Marles cam-and-lever steering box, with a ratio of 26:1. There was no power assistance.

Steering action from the wheel was transmitted via a short relay shaft, or drag link, to the offside front wheel, and then by means of a tie rod to the nearside. The steering wheel was 533mm diameter, giving approximately 4.5 turns from lock to lock for a turning circle of 16.8m.

Braking system
The four-wheel brakes were actuated by a conventional foot pedal, and were either of the air-assisted hydraulic type (early vehicles only), or were compressed air-operated; vacuum-operated brakes were fitted to post-war Air Ministry vehicles.

Air pressure was generated by a twin-cylinder compressor, belt-driven from the gearbox, and maintained in either one or two chassis-mounted reservoirs. The braking system air was drawn through the engine, with a gauge on the dash indicating braking system air pressure to the driver. Twin airline couplings were provided at the rear to allow connection to a trailer braking system.

The brake drums were of cast-iron, and braking effort was supplied by 19mm thick friction linings, 432mm diameter x 75mm width at the front, and 432x150mm at the rear, giving a total lining area of $0.310m^2$.

The end of the road - this Matador awaits disposal (KP)

Another Matador artillery tractor lines up for the auction (KP)

One of 10 low-silhouette Matador 6x6 artillery tractors, 1944/45 (IWM)

Some vehicles were also fitted with Warner electric brake units for use with trailers having electrically-operated brakes. In operation, the units allowed electric current at 6V, to be fed to the trailer brakes, in direct proportion to the brake pedal position, and the amount of air admitted into the control unit. The controller applied the trailer brakes slightly ahead of the tractor brakes.

The hand (parking) brake operated on the normal brake shoes on the rear wheels only by a system of mechanical levers; the actuating lever in the cab was operated by a pawl and ratchet.

Road wheels
The road wheels were 10.00x20in, shod with 14.00x20 bar-grip pattern cross-country tread tyres. Non-skid chains could be fitted, front and rear.

A single spare wheel was carried inside the rear body, and the braking system compressor could also be used for tyre inflation.

Chassis
The alloy-steel chassis was of bolted construction, as was normal AEC practice. The main members were formed from 204x76x6.4mm pressed channel sections. A laminated spring cross member was provided at front and rear to provide a support for the tow hitch; War Office and Admiralty contract vehicles were provided with a locking hook-type pintle, while the Air Ministry vehicles employed a hook at the front and a simple towing pin at the rear.

There were no separate bumpers.

Cab and bodywork
Cab
All of the 'Matadors' were of the forward-control pattern, with the cab installed over the engine on flexible mountings.

The cab was a relatively-crude affair, consisting of virtually-flat steel panels fitted to a timber frame. The commercial-style steel roof, with a lower profile and compound curves, was used on some of the early examples, and also on the 6x6 chassis it seems, but most 'Matadors' had a simple high curved roof of canvas on a wooden framework, giving a distinctive arched appearance. A steel or canvas-covered roof-mounted observation hatch was provided above the passenger seat.

Flat, curved mudguards, with dual reinforcing swages, were installed at the front, bolted up into the cab wheel arches. Some vehicles were fitted with just one headlight, others had two. Very early WW2 vehicles were fitted with a horizontal metal plate, located just below the driver's screen and painted with gas detector paint, but this was later deleted from production and removed from vehicles which had been so equipped.

There was a traditional cast radiator installed on the flat front panel.

The windscreen was of the standard four-piece divided design used on all military vehicles at the time, with the upper glazed lights hinged at the top edge. Large, fixed quarterlights were installed forward of the doors at a slight angle from the cab sides, and there were drop-down windows in the doors themselves.

The cab included seating for two using upholstered bucket seats, adjustable for height and reach; the passenger's seat could be folded to form a standing platform for use with the observation hatch.

Artillery tractor bodywork
The composite wood-and-steel rear bodywork of the artillery tractor was provided with two hinged doors in the body sides at the forward end. The side panels were fixed, and there was a hinged tailgate. Seating was fitted in the rear body for nine or ten passengers, together with stowage for their personal kit. A simple pressed-steel mudguard assembly covered the rear wheels.

There were shell carriers attached to rails in the floor, intended to make it easy to slide the rounds towards the tailgate, and thus offload to the gun itself.

Like the cargo versions, a removable canvas cover was supported on steel hoops to cover the cargo area; on early vehicles, the canvas cover was sometimes partially replaced by a metal top panel.

Winch
Most vehicles were fitted with a winch, either a 5 or 7 ton Turner horizontal-type machine, mounted on trunnion blocks between the chassis side members, midway along the length of the chassis.

The drum was mechanically-driven by a power take-off and shaft provided on the auxiliary gearbox, and was engaged by means of a dog clutch operated from inside the cab. A safety device was fitted which stopped the engine when the pull on the winch cable approached the maximum load rating. Pulleys and fairleads were fitted to allow forward or rear-ward winching. The winch cable was 17.5mm diameter, with a length of 63.25m; average cable speed was 16.4m/min in first gear with the engine running at 1000rpm, and 26.85m/min in second gear.

Electrical equipment
All models were wired on a balanced three-wire 12V system using an insulated return. The power was derived from four 6V 150Ah batteries which were wired in series-

parallel to provide 24V for starting. The lighting and accessory circuits were distributed on each side of the three-wire system (+12V/0/-12V) to provide an approximately-equal load on each pair of 6V batteries.

The starter and generator were standard commercial types; the generator had a rated output of 20A at a nominal 24V, under compensated voltage control.

DOCUMENTATION

Technical publications
User handbook
Tractor, GS, 10 ton, 4x4, AEC O853. WO code 17796.

Servicing schedule
Tractor, GS, 10 ton, 4x4, AEC O853. WO code 10810.

Parts list
Tractor, GS, 10 ton, 4x4, AEC O853. WO code 17670.

Technical handbook:
Waterproofing instructions: tractor, GS, 10 ton, 4x4, AEC O853. WO code 16526.

Tools and equipment
Tables of tools and equipment for 'B' vehicles: tractor, GS, 10 ton, 4x4, AEC O853. WO code 17722, Table 1086.

CHAPTER 1.2

AEC MILITANT FV11000 SERIES
FV11001, FV11002, FV11041, FV11044

During WW2, AEC had supplied more than 10,000 4x4, and latterly 6x6, 'Matadors' for use in a variety of roles, including cargo trucks, artillery tractors, tankers, and mobile cranes. Although production of the 'Matador' had nominally ceased in 1945, many of the vehicles remained in service throughout the late 'forties and early 'fifties. Obviously the War Office felt that the vehicles were giving satisfactory service, for in a somewhat unusual move, between 1953 and 1959, a further batch of 1800 'Matadors' (see Chapter 1.1) was purchased.

However, it was clear the 'Matador' could not go on for ever, and in 1952, an order was placed with AEC for some 3200 examples of a new vehicle, dubbed 'Militant', and intended to replace the by-then ageing WW2 'Matadors'.

The 10 ton Mk 1 'Militant' chassis was produced in both 6x4 and 6x6 configurations for use as a medium/heavy artillery tractor, designated FV11001/FV11002. The chassis was also used as a three-way tipper (FV11005), 10 ton cargo vehicle in 14ft (FV11007) and 18ft (FV11008) versions, and a crane (FV11009).

The 'Militant' was a thoroughly-conventional design produced from largely-commercial components, and it incorporated none of the innovations which were being implemented in the mainstream of post-war FVRDE vehicle design. In fact, the Militant' was quite recognisably based on the WW2 'Matador', so it should be no surprise that by 1959, both the War Office and FVRDE had begun to suggest that the 'Militant' Mk 1 had also become something of an antique, and were discussing a more modern successor.

In 1962, a Mk 2 vehicle was produced (FV11041) as a prototype replacement for the medium/heavy artillery tractor but there was no volume production. It was not until 1965, and the advent of the altogether more-modern Mk 3 'Militant', that the rather-archaic 'Matador' and early 'Militant' cargo vehicles were replaced.

The Mk 3 'Militant' was produced with a cargo body (FV11047), and as an armoured command vehicle (FV11061), but it is only the 10 ton 6x6 medium recovery vehicle (FV11044) that falls into the scope of this book.

With recovery equipment produced by Thornycroft, by then a long-time member of the ACV Group, the GS-classified FV11044 was designed as a replacement for the Scammell 'Explorer' (see Chapter 5.3) which by that time was getting a little long in the tooth. The FV11044 was intended to work alongside the CT-rated Leyland

TUGS OF WAR   19

CHAPTER 1.2: AEC MILITANT FV11000 SERIES

'Martian' 10 ton recovery vehicle (see Chapter 4.2). Unlike the Scammell which was equipped only with a mechanical winch for the recovery crane, the 'Militant' was fitted with a hydraulic telescopic crane, similar to that used on the Leyland, and prototype chassis of the 'Militant' were also produced with the soon-to-be fashionable multi-fuel engine.

In all, some 200 examples were purchased, entering service in 1971, with some remaining in use until well into the 'nineties. It was only the advent of the more-sophisticated EKA hydraulic-equipped recovery vehicles that sounded the death knell for the 'Militants'.

DEVELOPMENT

In 1951, AEC started work on a new 10 ton chassis, designed to replace the WW2 'Matador'. The new chassis was based on the rigid 6x6 produced for the RAF during WW2, which was a hybrid of the existing 'Matador' and 'Marshal' models.

The new vehicle was dubbed 'Militant', a name originally associated with the Maudslay Company, which had been absorbed into the ACV Group in 1948. Maudslay had produced a 6 ton 4x2 truck named 'Militant' during WW2, but it was the ageing AEC 'Matador', rather than the 'Militant' which its post-war namesake was intended to replace, and despite the new name, the vehicle shared many design principles with its WW2 'Matador' predecessor.

The chassis was given the AEC type number O859 in 6x4 form, and O860 in 6x6 form.

The 'Militant' was intended to take its place among the GS classified vehicles which formed part of the FVRDE-designed post-war vehicle plan. This plan, which had originated in the late 'forties, divided vehicles into three classes. Firstly, came the CT, or combat class, consisting of purpose-designed, highly specialised military vehicles, intended for 'front line' roles. Next, were the general-service, or GS, class - these were largely commercial designs, but were adapted for military service by the use of all-wheel drive, larger wheels and tyres, and standardised accessories. Finally, there was the commercial, or CL class, consisting of virtually-standard commercial vehicles.

Within each weight category, there was usually an equivalent CT and GS vehicle, and often a CL vehicle as well. In the 10 ton category, the CT vehicle was the Leyland 'Martian'; while the AEC 'Militant' was its GS counterpart.

Although the FVRDE specification for the cab was not finalised until May 1956, deliveries of the 'Militant'

VEHICLE OUTLINES

FV11001/FV11002

FV11044

SCALE 1:100

SPECIFICATION

Dimensions and weight

	FV: 11001	11002	11041	11044
Dimensions (mm):				
Length	7353	7353	7470	8230
Width overall	2438	2438	2440	2500
Height:				
to top of cab	2807	2807	2807	3020
maximum height	2946	2946	3040	3100
load platform	1370	1370	1370	-
Wheelbase	3924	3924	3924	3924
rear bogie	1372	1372	1372	1372
Track:				
front	2000	2000	1990	2000
rear	2235	2235	1900	1990
Ground clearance	336	336	336	400
Weight (kg):				
Laden	15,425	15,710	15,710	21,040
Unladen	10,325	10,580	10,580	18,289
Bridge classification	16	16	16	20
Crane dimensions (mm):				
Minimum reach	-	-	-	3124
Maximum reach	-	-	-	5563
Maximum lift height	-	-	-	3048
Maximum slew, left/right	-	-	-	120°
Max load (kg)	-	-	-	5250
Load at max reach (kg)	-	-	-	2642

Performance

	11001	11002	11041	11044
Speed (km/h):				
maximum, on road	39	39	42	78
average, off road	19	19	19	25
Fuel consumption (l/km)	0.47	0.47	0.47	0.36
Maximum range (km)	480	480	480	480
Turning circle (m)	17.69	17.9	17.9	23.8
Maximum gradient	33%	33%	-	<50%
stop and restart	33%	33%	-	50%
Approach angle	-	-	-	37°
Departure angle	-	-	-	32°
Fording (mm):				
unprepared	760	760	760	760
prepared	1980	1980	1980	1980
Capacity (kg):				
Maximum load	5130	5130	5130	-
towed load	16,256	16,256	16,256	10,181

began in 1952. The vehicle was available in 6x4 and 6x6 configurations, but only the artillery tractor employed the 6x6 drive line. The engine was the well-proven AEC type A223 six-cylinder diesel engine of 11.3 litres.

The 'Militant' was originally conceived as a cargo truck, but other variants were planned, and in about 1954, an artillery tractor was produced. This was intended for use with medium and heavy field artillery pieces up to 16 tons in weight, for example, the 40mm Bofors anti-aircraft gun, and the 5.5in howitzer; it was also intended that the FV11002 'Militant' be used to tow the 7.2in howitzer when the WW2 Mack tractors were replaced. However, the vehicles were always in such short supply that in the end the FV1103 Leyland was assigned this role.

The wood-and-steel composite body, constructed by Crossley Motors, was designed to accommodate a crew of nine, plus two in the cab, together with stowage for their personal kit, as well as up to 4500kg of stores and ammunition.

The vehicle was able to tow a gross weight of 16,256kg under cross-country conditions, and offered excellent off-road performance through the use of a fully-articulated, centrally-pivoted rear bogie. This configuration allowed up to 305mm relative movement of the wheels on diagonally-opposite sides of the bogie.

A total of 3200 Mk 1 'Militants' was produced, with 349 supplied for the artillery tractor role; these were ordered under four contracts (6/Veh/8238/CB27a, 15035, 23295 and 26136). Production ceased in 1964, and although the vehicle served into the 'seventies, before the end of the 'sixties FVRDE had begun to examine the possibility of an improved version.

Mk 2 'Militant'

As early as 1959, the War Office was beginning to acknowledge that the Mk 1 'Militant' was of 'obsolescent design'. At a meeting convened at Chertsey in February of that year, some 16 representatives of FVRDE, the War Office, and the user arms got together to discuss AEC's proposals to update the vehicle. It is interesting to note that, among others, the FVRDE team included Rex Sewell, the designer responsible for much of the work on the Austin FV1800 Series, better known as the 'Champ'.

War Office Policy Statement (WOPS) 71 had set out the desirability of adopting multi-fuel engines, and AEC proposed that the standard 11.3 litre diesel engine be replaced either by an AEC AV690 diesel engine, or by the 2AV760 multi-fuel engine which at that time was still under development. Other changes included the replacement of the main and auxiliary gearboxes by heavier-duty units, better suited to continuous working in low gear. Stability was to be improved by the use of a wider track. The cab was also redesigned, following commercial practice, to improve crew comfort and to provide a more-modern appearance. Finally, there were a large number of detail changes, either to conform to latest commercial practice, or to fall into line with developments at FVRDE.

There was some discussion regarding the problems of providing adequate stowage for the crew's kit in the shallow, commercial cab design. It was agreed that investigations would be conducted into the possibility of providing external, waterproof stowage behind the cab, or of providing internal stowage on top of the engine cowling.

Apparently, the changes mooted by AEC would result in an overall saving per chassis in the order of £100, and it was decided that, alongside the Mk 2 cargo vehicles, a total of 10 artillery tractors would be ordered, complete with composite wood-and-steel Park Royal bodies. Designated FV11041, two of these were for FVRDE trials, while the remainder were to be put through a 12-month user trial programme, starting no later than September 1959.

It was planned that all 'Militants' purchased for 1961/62 would be of the Mk 2 type, and AEC had stated that design clearance would be required by September 1960 if deliveries were to be on time.

As it happened, there was no volume production of the Mk 2, and although examples of the artillery tractor were exhibited at the 1962 FVRDE exhibition of military vehicles at Chertsey, only the 10 prototypes were produced. The multi-fuel engine was not available in time to be tested in the Mk 2 chassis, at least not before the Mk 3 was under development, and it was not until the advent of the Mk 3 'Militant' that this particular development got off the drawing board at all.

A handful of the prototype Mk 2 vehicles remained in service during the 'sixties, but there was no further discussion regarding the placing of contracts. In the report which accompanied the Mk 3, AEC themselves admitted that events had really overtaken the Mk 2 and that, in fact, there would never be any series production.

Mk 3 'Militant'
Arising out of the experience gained with the trials of the Mk 2 prototypes, in March 1963, FVRDE issued a specification (no 9297) for a further-improved 'Militant' chassis, which was ultimately to enter production as the Mk 3. In July of that same year, AEC submitted a formal proposal to FVRDE which described a 10 ton 6x6 chassis-cab, developed in consultation with the FVRDE design

Prototype or development Mk 1 Militant with simple ballast body (IWM)

Development vehicle from rear (IWM)

FV11002 6x6 Mk 1 Militant artillery tractor (IWM)

CHAPTER 1.2: AEC MILITANT FV11000 SERIES

Mk 1 Militant artillery tractor viewed from above showing seating (IWM)

Mk 1 Militant cargo vehicle used identical cab (IWM)

FVRDE chassis shot of 6x6 Mk 1 Militant (IWM)

team, and designated AEC type O870. AEC believed that this vehicle satisfied FVRDE Specification 9297.

The Mk 3 'Militant' was a far more modern truck, and unlike the abandoned Mk 2 project was more than simply a warmed-over version of the earlier vehicle. Aside from the multi-fuel engine which was still under development, the chassis was based on proven commercial components, and was offered in 3924 and 4880mm wheelbase lengths. Variants proposed by AEC, and presumably by FVRDE, included artillery tractor, recovery tractor, fifth-wheel tractor, cargo truck, tipper, and fuel tanker.

AEC pointed out to the Ministry that production would be split between the companies within the ACV Group. Engines and chassis were to be produced at AEC's Southall factory. Thornycroft, who were still based at Basingstoke, would produce the gearboxes and the recovery equipment. And finally, the Warwickshire-based Maudslay Motor Co were to be responsible for the axles. Final assembly was to take place at the AEC works but the major advantage of this arrangement was that, should there be a sudden increase in demand, each of the three production facilities could increase production with the minimum lead time.

Research work on multi-fuel engines, able to run on petrol, diesel, aviation spirit, and apparently even sump oil, had started in Germany in the early 'fifties. It was intended from the start that the Mk 3 'Militant' would use a multi-fuel engine. For the prototypes, AEC offered a reworked AV690 engine, and a new unit, designated AV760 which was likely to be available by the end of 1964. Both were designed for multi-fuel operation, even though the AV760 was available initially in mono-fuel form only.

The transmission consisted of a six-speed, constant-mesh main gearbox, with a two-speed auxiliary gearbox and transfer case mounted directly to it. Suspension was by semi-elliptical springs, with the rear axles trunnion-mounted to allow maximum articulation; both front and rear axles were of the double-reduction type. A shallow cab was used on the cargo vehicles, similar to that fitted to the prototype Mk 2 chassis.

At the end of 1964, at a meeting between FVRDE and AEC, it was agreed that the first Mk 3 'Militant' prototype would be delivered in December of that year, and would be fitted with the newly-developed AEC AV760 engine. Before that time, a Mk 2 tractor fitted with an AV690 commercial diesel engine, modified to allow it to run on alternative fuels, and thus re-designated 2AV690, was also to be delivered to FVRDE. Both vehicles were to be subjected to road testing of 16,000km duration with the intention of completing the tests by May 1965. Subject to some agreement on engines, it was decided that six chassis would be produced, and that user trials for the chassis, and the proposed power units, would begin in March/April the same year, with production scheduled to start in late 1966. If the trials were successful, AEC agreed that up to 15 Mk 3 machines could be supplied in lieu of a similar number of Mk 1's which were already on order.

AEC had also agreed that development of the multi-fuel engines would be at their own expense, and a separate team was set up for this work, led by W A Edmonds from FVRDE, with C K Martlew and C Marsh from AEC. Three of the new type AV760 multi-fuel engines were constructed during 1964, and during the trials period, one of the Mk 2 chassis was used as a mobile test bed for both the type AV690 mono-fuel engines, and AV760 multi-fuel units, while the Mk 3 chassis were fitted with either the 2AV690 or AV760 multi-fuel units.

By 1965 it was obvious that development of the multi-fuel engine was not going particularly well.

One of the difficulties with multi-fuel engines is that of providing adequate lubrication for the upper valve gear. While DERV has natural lubricating properties, petrol does not, and in the early days of development this often led to overheating and failure when running on petrol. During road trials, the AV760 engine seized after just 948 miles on petrol; the 2AV690 managed 5294 miles on petrol before the trial was terminated due to misfiring. Both engines suffered from severe piston erosion, presumably due to uncontrolled fuel detonation. New flat-topped pistons were produced to replace those with a built-in combustion chamber, raising the compression ratio to 19:1, and presumably solving the erosion problem. However, other problems remained, and in July 1965 it was stated the 2AV760 engine 'had not shown itself to be satisfactory in service, and could not be accepted at the present time'.

Obviously the development of a satisfactory multi-fuel engine was going to be a somewhat longer-term business than had originally been envisaged, and by the time the Mk 3 was ready for delivery, the multi-fuel engines were still not operational. The specification for the prototype vehicles was amended to revert to the AV760 commercial diesel engine, with a compression ratio of 16:1.

In due course, attention was turned to the FV11044 recovery tractor. In early 1965, it had been agreed that four prototypes would be required for the recovery vehicle variant. These were to be made available between October 1965 and January 1966, with a minimum of 300 chassis required once full production was underway. As it happened, prototype trials did not begin until 1967

FV11041 Mk 2 Militant artillery tractor (TMB)

Fine view of the Mk 3 Militant chassis (BCVM)

CHAPTER 1.2: AEC MILITANT FV11000 SERIES

FV11044 Mk 3 Militant heavy recovery vehicle (REME)

Mk 3 Militant chassis from rear showing axle articulation (BCVM)

Vertical Scammell winch used on Mk 3 Militant chassis (BCVM)

and were continued through 1969. The basic chassis was unchanged from the cargo truck but a deeper cab was employed which provided seating for three, together with additional stowage space. The hydraulic recovery equipment was supplied by Thornycroft, and the crane itself was manufactured by the British Crane and Excavator Corporation, or Coles, with the winch supplied by Scammell.

Mk 3 vehicles finally began to enter service in 1966, with some 200 recovery tractors being delivered from about 1971.

NOMENCLATURE and VARIATIONS

FV11001. Tractor, 10 ton, 6x4, GS, medium/heavy artillery, AEC

Three-axle four-wheel drive artillery tractor with two-man cab and composite wood-and-steel rear body, produced by Crossley Motors, and designed for a crew of seven, together with 4500kg of stores and ammunition. Equipped with AEC A223 11.3 litre direct-injection diesel engine, five-speed gearbox and two-speed transfer case. Fitted with 5 or 7 ton mechanical winch.

FV11002. Tractor, 10 ton, 6x6, GS, medium/heavy artillery, AEC

Six-wheel drive development of FV11001 tractor; all other details similar.

FV11041. Tractor, 10 ton, 6x6, GS, medium/heavy artillery, AEC Mk 2

Interim development of the FV11002 artillery tractor, with similar rear body, in this case manufactured by Park Royal Vehicles, and new cab similar to that ultimately used on the Mk 3 chassis. Fitted with AV690 11.3 litre diesel engine; otherwise details largely as FV11001/FV11002.

Used as a test bed for the Mk 3 vehicle and for the projected multi-fuel engines; no series production.

FV11044. Tractor, 16 ton, 6x6, GS, heavy recovery, AEC Mk 3

Heavy recovery tractor intended to replace the FV11301 Scammell 'Explorer'. Suitable for recovery, and transport on suspended tow, of all wheeled vehicles up to the 10 tonne class; fitted with hydraulic power-operated crane, and 15 tonne mechanical winch. New, three-man cab, available in both left- and right-hand drive configurations.

Powered by an AEC AV760 diesel engine of 12.47 litre, driving through a six-speed gearbox and two-speed transfer case; driven axles front and rear, with four differentials, and inter-axial differential locks.

TRAILERS

The FV11044 recovery tractor was intended to be used occasionally with the 10 ton light recovery trailer, which had been designed for recovering and transporting CT and GS category wheeled vehicles up to 10 tons in weight, as well as being suitable for carrying the FV430 Series tracked vehicles:

FV3221. Trailer, 10 ton, 4TW/2LB, recovery, light; manufactured by Crossley Motors Ltd, Rubery Owen & Co Ltd, and J Brockhouse & Co Ltd.

From time to time, the artillery tractors might also have been found towing equipment trailers.

DESCRIPTION

Engine

In all, five different engines were fitted into the 'Militant' chassis during its life, all produced by AEC. The A223, AV690 and AV760 diesels were used in the production examples, while the developmental 2AV690 and 2AV760 multi-fuel engines were trialled in both the Mk 2 and Mk 3 chassis.

Mk 1 'Militants' were fitted with AEC's familiar A223 direct-injection, water-cooled four-stroke diesel engine, producing a maximum of 150bhp from 11.3 litres, at a governed speed of just 1800rpm. The engine was an in-line six, with overhead valve gear driven by pushrods from a low-mounted camshaft, and with CAV fuel-injection equipment. The block was a two-piece casting, with the upper portion of fine-grained iron, fitted with centrifugally-cast dry cylinder liners; the lower part was of aluminium alloy. The twin cylinder heads, each covering three cylinders, were also of fine-grained cast iron.

The engine planned for the production Mk 2 vehicles was a development of the earlier unit, designated AV690. The 'V' addition to the code indicated that this particular engine configuration was intended for 'vertical' applications; there were also 'H' coded units (AH690) intended for horizontal use, though none were employed in 'Militants'. The capacity remained at 11.3 litres but the power output was up to 192bhp (gross) at 2000rpm.

The Mk 2 was also used as a test bed for the development of the multi-fuel engines, designed to run on either petrol or diesel fuels, but in both cases as a compression-ignition unit. The engines employed were a multi-fuel version of the existing unit, known as 2AV690, which produced 170bhp at 2000rpm (running on diesel), as well as a completely-new design known as 2AV760.

The Mk 3 was prototyped with a mono-fuel (diesel) version of the AV760, as well as with the 2AV690 multi-fuel engine, but all production vehicles were fitted with the AV760 engine running on diesel only. This engine was also an in-line, four-stroke water-cooled six, with a one-piece cast-iron crankcase fitted with renewable dry cylinder liners; valves were overhead, operated by conventional pushrods. In use, the unit produced 226bhp at 2200rpm, from a cylinder capacity of 12.47 litres.

Engine data

	A223	AV690	AV76
Capacity (cc)	11,310	11,310	12,473
Bore and stroke (mm)	130x142	130x142	136x142
Compression ratio	16:1	16:1	16:1
Power output (bhp):			
gross	150	192	226
net	144	170	204
Maximum torque (lbs/ft):			
gross	510	557	588
net	480	500	560
Firing order	153624	153624	153624

Cooling system

All of the AEC engines were water-cooled in the conventional manner using a pump-assisted, pressurised ($7050 kgf/m^2$) thermo-siphon system in conjunction with a 'Still' tube radiator on Mk 1 chassis, and a film block radiator on Mk 2 and Mk 3 models.

A large diameter multi-bladed fan, 610mm on Mk 1 vehicles, 560mm on Mk 3, was used to draw air through the radiator.

On Mk 1 tractors, a tubular oil cooler was mounted in front of the radiator; on Mk 3 models, the oil cooler was of the air/water type.

Transmission

On Mk 1 and Mk 2 vehicles the power was transmitted through a 400mm single dry-plate clutch, to a five-speed and reverse gearbox, unit-constructed with the engine. First and reverse were engaged by a sliding gear, whilst the other speeds were of the constant-mesh helical type.

There was a separately-mounted, two-speed auxiliary gearbox, which also contained the transfer gear assembly for 6x6 vehicles; the front-wheel drive was interlocked with the low ratio of the auxiliary gearbox, and the auxiliary gearbox also provided drive for the winch.

The Mk 3 'Militant' was fitted with a completely-new six-speed transmission which incorporated an overdrive gear. A two-speed auxiliary gearbox and transfer system was mounted directly on the back of the main gearbox.

Front-wheel drive could be engaged with either the high or low ratios. The clutch was of AEC's own design, with the diameter increased to 432mm, and the release system changed from mechanical to hydraulic.

The transfer box to axle ratios were 1:1 and 1.62:1 on Mk 1 and 2 vehicles, or 1:1 and 1.87:1 on the Mk 3. The winch drive ratios were 1.62:1 or 1:1, on Mk 1/Mk 2, and Mk 3 vehicles respectively.

From the auxiliary gearbox, an open propeller shaft ran the length of the vehicle to drive the first of the rear axles, with a second, shorter shaft running from this to the second rear axle. On Mk 1 6x6 vehicles, and on Mk 2 and Mk 3 vehicles, a third propeller shaft ran from the transfer case to the front axle.

Suspension and axles
On all three models, the front suspension was by means of long semi-elliptical multi-leaf springs, fitted with double-acting telescopic shock absorbers.

At the rear, the fully-articulated bogie was suspended on inverted springs pivoted on a cross tube mounted in cast brackets. On the Mk 1 and Mk 2 vehicles, the spring ends were formed into an eye which was attached by means of a pin and bush to the axle casing; on Mk 3 models, the outer ends of the springs remained free to slide in axle-mounted stirrups. Torque reaction was transmitted to the chassis by means of rubber-jointed parallel rods.

A mechanical locking device was provided on the rear suspension of the recovery vehicle to assist in lifting operations.

Drive to the rear axles was transmitted through a double-reduction spiral-bevel/double helical-gear type differential within each axle casing, and then through fully-floating axle shafts to the four road wheels, giving an overall axle ratio of 7.9:1. While AEC had made some play of the fact that 'to provide maximum traction under the most severe conditions' there was no inter-axial differential on the Mk 1 vehicles, on the Mk 3 chassis they obviously had second thoughts about this, and a third, inter-axial lockable differential was housed in the centre axle casing.

On Mk 1 6x6 vehicles, and Mk 2 vehicles, the front axle consisted of a one-piece solid forging with drilled axle tubes; on Mk 3 vehicles, the axle casing was of fabricated construction. Both types of driven axle incorporated double-reduction spiral-bevel/double-helical gear drive, with Bendix 'Tracta' type constant-velocity joints transmitting power to the hubs via fully-floating axle shafts. In both cases, the axle ratio was again 7.9:1.

On 6x4 chassis, the front axle was a simple 'I' section stamping, of the reversed Elliott type; thrust from the swivel pins was taken by hardened-steel thrust buttons.

Steering gear
Steering was controlled through an AEC-designed worm-and-nut steering box; the gear ratio was 32:1 on Mk 1 and Mk 2 vehicles, 40:1 on Mk 3. Although power-assisted steering was offered on Mk 1 vehicles, it was not fitted on chassis supplied to the British Army, but all of the Mk 3 vehicles were fitted with hydraulic assistance.

Steering action from the wheel was transmitted via a short relay shaft, or drag link, to the offside wheel station, and then by means of a tie rod to the nearside. The hydraulic assistance on Mk 3 models was provided by means of a double-acting ram fitted between the chassis and the drop arm, with the drag link connected to the end of the ram. The reaction valve was installed in unit with the ram.

Mk 3 chassis were available fitted for both right- and left-hand steering.

The steering wheel was 533mm diameter, giving approximately six turns from lock to lock.

Braking system
All six wheels were provided with single leading shoe cam-expanded, air-operated brakes, actuated by a conventional foot pedal. In addition, there was a hand-operated trailer braking system. Detail changes were made to the braking system of Mk 3 vehicles following chassis number 21165.

Air pressure was generated by a twin-cylinder compressor, belt-driven from the rear of the auxiliary gearbox, and maintained in two chassis-mounted reservoirs. The braking system air was drawn through a combined air cleaner and anti-freeze device, designed to prevent the braking system from icing up. A gauge on the dash indicated braking system air pressure to the driver. Twin airline couplings (triple on Mk 3) were provided at the front and rear to allow connection to the braking system of a trailer or disabled vehicle under tow.

The brake drums were of cast-iron and braking effort was supplied by 19mm thick friction linings. On the Mk 1 and Mk 2 vehicles, the brake lining sizes were 425mm diameter x 75mm width at the front, and 388x162mm at the rear, giving a total lining area of $0.578m^2$. Larger drums were employed on the Mk 3 chassis: 432x102mm on the front wheels, and 394x108mm at the rear, giving a braking area of $0.7746m^2$.

All Mk 1 artillery tractors were fitted with Warner electric brake units for use with trailers having electrically-operated brakes. In operation, the units allowed electric current at 6V, to be fed to the trailer brakes, in direct proportion to the brake pedal position, and thus the amount of air admitted into the control unit. The controller applied the trailer brakes slightly ahead of the tractor brakes. An FVRDE memo dated 11 February 1959 stated that this system would not be required on the Mk 2 (and by implication Mk 3) vehicles. However, the Mk 3 recovery tractor was provided with an additional pressure-operated hill-holder braking system operating on the front axle, the second of the rear bogie axles, and on the trailer.

On Mk 1 vehicles, the hand (parking) brake operated on the normal brake shoes on the rear four wheels only by means of rods and levers. A conventional ratchet-operated lever was provided in the cab. Operation of the hand brake system on Mk 3 chassis was air-assisted.

Road wheels
For all models, the road wheels were 10.00x20in, divided-disc type on Mk 1 and Mk 2 vehicles, and four-piece disc type on Mk 3. Tyres were 15.00x20 bar-grip pattern cross-country tread on Mk 1 chassis; 14.00 or 15.00x20 on the Mk 2; and 16.00x20 Michelin XL pattern radials on the Mk 3. Non-skid chains could be fitted to the front wheels, and overall track-type chains to the rear.

On Mk 1/Mk 2 tractors, a single spare wheel was carried vertically behind the cab in a hinged frame designed to simplify handling; on Mk 1 vehicles, the frame was described as being of the 'Kennedy & Kempe' pattern, while on the Mk 2, the frame was produced by AEC themselves. The Mk 3 recovery vehicles carried a single spare wheel, carried flat on the rear deck positioned for easy loading and unloading by the crane jib.

The braking system compressor could also be used for tyre inflation.

Chassis
In accordance with standard AEC practice, the chassis was of bolted construction. The main chassis members were formed from 254x76x8mm pressed channel sections, with an 8mm thick flitch fitted over the rear bogie; channel section cross-members were installed to provide rigidity, with an 'I' section cross-member at the engine front support.

Both models were fitted with rotating towing pintles on the rear cross member; a similar hitch was also fitted at the front of the Mk 3 model, while the Mk 1 employed a simple towing pin in this position. A full width front bumper was used, with corner bumper plates at the rear; the front bumper on Mk 3 models was shaped to allow the tow hitch to be inset, thus providing an unobstructed bumping surface.

Cab and bodywork
Cab
All of the 'Militants' were of the forward-control pattern, with the cab installed over the engine on flexible mountings. The Mk 1 models used a purpose-designed, deep-pattern military cab with a traditional cast radiator installed on a shaped front panel, while the Mk 2 and Mk 3 vehicles used a modified commercial cab with a simple wire-mesh radiator grille installed on the cab front. Small semi-enclosed mudguards were installed at the front, bolted up into the cab wheel arches.

The cab was constructed from double-skinned steel panels, with the cavity packed with insulating material. A joint was provided at the waist rail to permit the overall height to be reduced for shipping, and a roof-mounted observation hatch was provided.

A two-piece, laminated windscreen was provided, hinged at the top edge on Mk 1 vehicles, and fixed on Mk 2 and Mk 3's. Quarterlights were installed forward of, and behind the doors, and the doors themselves were fitted with drop-down windows.

The Mk 1 cab included seating for two using upholstered bucket seats, adjustable for height and reach; the passenger's seat could be folded to form a standing platform for use with the observation hatch. There was seating for three in the Mk 3 cab.

Artillery tractor bodywork
The composite wood-and-steel rear bodywork of the artillery tractor was constructed by Crossley Motors, and was provided with two hinged doors in the body sides at the forward end. The side panels were fixed, and there was a hinged tailgate. Seating was fitted in the rear body for nine passengers, together with stowage for their personal kit. A simple pressed-steel mudguard assembly covered both rear wheels, and there appear to have been several different patterns fitted.

Like the cargo versions, a removable canvas cover was supported on four steel hoops to cover the cargo area.

Recovery vehicle equipment
The recovery equipment was mounted on a subframe of supporting channels, to which was mounted a simple steel platform body. Stowage lockers in the platform were provided for the various items of loose equipment.

Two jib support legs were fitted to allow suspended tows from the jib, while an 'A' frame bracket was bolted to the rear support channels; the 'A' frame could be fitted in one of three positions. Four outrigger jacks were attached to the subframe at strategic points to provide stability during recovery operations, and a hydraulically-operated spade-type earth anchor at the rear of the vehicle allowed pulls up to 30 tons.

A dual-section extensible jib, mounted on a vertically-pivoted post, was bolted to the recovery equipment subframe. Jib extension, luffing and slewing actions were provided by means of hydraulic rams. Hoisting was effected by a hydraulic capstan winch mounted at the end of the jib, with a disc brake to hold the load. The crane was suitable for both workshop and recovery operations, and jib and winch control was carried out from an operator's chair on the right-hand side of the outer column. A covering canopy was carried as loose equipment.

A cable-operated 'reactor' device, sometimes called a back-acter, was fitted to prevent undue weight transference from the front wheels during suspended towing operations. This could be considered a technological solution to the problem which, on the Scammell 'Pioneer' was solved by hanging several hundredweight of pig iron ballast over the front wheels.

Winch

Two types of winch were employed according to mark; a Turner 5 or 7 ton winch was fitted on Mk 1 vehicles, while the Mk 3 models used a 15 ton Scammell winch.

The Turner winch was mounted between the main chassis members, directly behind the auxiliary gearbox. The vertical drum was mechanically-driven by means of a worm wheel, connected by chain to a power take-off shaft on the auxiliary gearbox shaft. A safety device was fitted which stopped the engine when the pull on the winch cable approached the maximum load rating. Pulleys and fairleads were installed to allow front or rear pulls; winching angles were 45° to the right or left at the front, and 30° to the right, or 90° to the left at the rear. The winch cable was 17.5mm diameter, with a length of 76.25m.

A Scammell horizontal winch was used on Mk 3 vehicles, installed between the chassis members immediately below the crane deck. The winch was driven by a power take-off on the auxiliary gearbox, operating by drive shafts through a dog clutch. The winch was arranged so as to permit front or rear pulls, up to 90° from the vehicle centreline at the rear and 20° at the front, by means of pulleys and rope fairleads. All winch controls were operated by air pressure, with the control levers installed in the driver's cab; an electrical cut-out circuit, wired through the ignition, was provided to protect the winch against overload. A 22mm diameter cable was used, with a total length of 137m; maximum winch speed was 30.5m/min.

Electrical equipment

As was the custom at the time, all models were wired on the 24V system, with negative return; the wiring on Mk 3 chassis used the double pole system. Mk 1 chassis were fitted with four 6V 110Ah batteries, installed behind the seats, while Mk 3 vehicles were fitted with two standard NATO 6TN 12V 100Ah batteries, installed centrally in the cab.

Simms or CAV starters and generators were fitted to the Mk 1 vehicles. The generator was an FVRDE-designed 'No 1, Mk 2/1' machine with a maximum output of 12A; while the starter was a 'No 2, Mk 1B' or 'Mk 1B/1' unit, offering axial engagement. By the time the Mk 3 was produced, dc generators had generally given way to higher-output alternators, and a CAV-produced 'No 10, Mk 2' three-phase machine was fitted, giving a maximum output of 90A; the starter was a 'No 2, Mk 1B/1' unit, or a 'No 5, Mk 1'.

DOCUMENTATION

Technical publications

Specifications
FVRDE Specification 9582: production specification, cab for truck, 10 ton, 6x4, and 6x6, GS, (AEC), FV11000 Series - various versions.
FVRDE Specification 9364: production specification, recovery vehicle, wheeled, medium, 6x6, AEC, Mk 3, FV11044.

User handbooks
Tractor, GS, 16 ton, HAA, 6x4, AEC O859. Army codes 17822, 17838.
Tractor, GS, 16 ton, HAA, 6x6, AEC O860. Army codes 17836, 18375.
Recovery vehicle, wheeled, medium, 6x6, AEC, Mk 3. Army code 22214.

Servicing schedules
Tractors, GS, 16 ton, HAA, 6x4 and 6x6, AEC O859/O860. Army code 10775.
Recovery vehicle, wheeled, medium, 6x6, AEC, Mk 3. Army code 60608.

Parts list
Tractors, GS, 16 ton, HAA, 6x4 and 6x6, AEC O859/O860. Army codes 17728, 17729.

Technical handbooks
Data summary: recovery vehicle, wheeled, medium, 6x6, AEC, Mk 3. EMER D120.

Technical description:
Tractors, GS, 16 ton, HAA, 6x4 and 6x6, AEC O859/O860. EMER D152.

Recovery vehicle, wheeled, medium, 6x6, AEC, Mk 3. EMER D122.

Modification instructions
Tractor, GS, 16 ton, HAA, 6x4, AEC O859. EMER D157, instructions 1-18.
Recovery vehicle, wheeled, medium, 6x6, AEC, Mk 3. EMER D127, instructions 1-13.

Standards
Field inspection standard: recovery vehicle, wheeled, medium, 6x6, AEC, Mk 3. EMER D128.
Base inspection standard: recovery vehicle, wheeled, medium, 6x6, AEC, Mk 3. EMER D128, Part 2.
Inspection standard: tractors, GS, 16 ton, HAA, 6x4 and 6x6, AEC O859/O860. EMER D158.

Miscellaneous instructions
Tractor, GS, 16 ton, HAA, 6x6, AEC O860. EMER D169, instructions 1-34.
Recovery vehicle, wheeled, medium, 6x6, AEC, Mk 3. EMER D129, instructions 1-8.

Complete equipment schedule
Recovery vehicle, wheeled, medium, 6x6, AEC, Mk 3. Army code 34144.

Tools and equipment
Tables of tools and equipment for 'B' vehicles: tractors, GS, 16 ton, HAA, 6x4 and 6x6, AEC O859/O860. Army code 18220, Table 1010.

ALBION
CHAPTER 2

CHAPTER 2: ALBION

For more than 70 years, the firm of Albion Motors was Scotland's largest producer of motor vehicles, with the first vehicle rolling out of the company's Glasgow factory in 1900.

The company was founded by Norman Fulton and T Blackwood Murray, who set up the embryonic Albion Motor Car Co in a 300m² factory at Finnieston Street, Glasgow in 1899. The first vehicles to be produced were dogcarts, but by 1902, the company had produced its first commercial vehicle, an 8hp half-ton van, which was really little more than a car chassis with a commercial body. At the same time, Albion moved to a new factory at Scotsoun, close to the River Clyde.

Although the Scottish market was not large, expansion was steady during the early years of the century. Agents were soon appointed in London to service customers the 'other side of the border', and the export trade was also proving brisk.

In 1910, Albion introduced their 32hp A10 chassis, which in both bus and truck form, was to go on to become one of the company's most popular models. More than 6000 A10 trucks were supplied to the War Office during the Great War, and production continued until 1929, by which time more than 9000 had been produced altogether. The military connection also continued during the early 'twenties with the SB24 1.5 ton 'Subsidy' model, which was among the first of such vehicles to be ratified under the War Office scheme.

The enormous advances in automotive technology which had taken place during the war and the immediate post-war years saw Albion, in common with many other manufacturers, introducing a multiplicity of new models during the 'twenties, each offering the operator improved performance and efficiency. The first forward-control (cab-over) model was introduced in 1924, with the first diesel in 1933. However, it had been possible for customers to specify, for example, Gardner or Beardmore diesels for some years before Albion introduced their own.

Although Albion had continued to supply vehicles to the War Office during the inter-war years, for example the AM463 heavy ambulance of 1934, and the BY1 3 ton trucks of the late 'thirties, the outbreak of World War 2 obviously brought about a considerable increase in orders.

Alongside the heavy CX 10 ton chassis which were employed as GS cargo trucks, artillery tractors and tank transporters, the company supplied a range of 3 ton 4x4 and 6x4 chassis for use as GS cargo vehicles, bridging units, machinery trucks etc, as well as quantities of the type AM463 2 ton chassis, originally supplied as an ambulance, with a refuelling body for the RAF.

In 1951, with a £3 million merger, Albion was absorbed into the growing Leyland empire. There was little overlap between the two companies' model ranges and for a while the merger seemed to make good economic sense, producing a much-strengthened company with a wide payload range. Albion continued to supply military vehicles during the late 'forties and early 'fifties, producing the WD66N chassis as a replacement for the ill-fated Vauxhall FV1300/FV14000 project. The Ministry of Supply also bought quantities of the WD/HD23 10 ton 6x6 chassis, bodied as mobile cranes, machinery trucks and tippers.

Despite the investment of some £2 million by Leyland in new equipment in the mid-fifties, expanding the Scotsoun factory to nearly 95,000m², the writing was probably already on the wall for Albion as an independent name, and by 1959, Albion trucks were beginning to adopt the standard Leyland look.

The Albion name survived until 1972 when the company was re-christened Leyland (Glasgow) Ltd, but from the early 'sixties, Albion had started to specialise in transmission components destined for use throughout the Leyland group. In 1981, Leyland's so-called 'Radical Plan' concentrated axle production at Albion, and vehicle production ceased altogether. The moving production line was dismantled and transferred to the Scammell plant at Watford, by then also a member of the mighty Leyland empire.

Axle-building continued at the old Albion factories throughout the 'difficult' Leyland years and even beyond the DAF takeover. In fact, after the merger, DAF showed great confidence in the Scottish workforce by investing in the plant with new equipment and facilities, and Albion remained the largest motor industry manufacturer in Scotland.

CHAPTER 2.1

ALBION CX and FT SERIES
CX22S, CX24S, CX33, FT15N/FT15NW

Albion had been established as a manufacturer of heavy commercial vehicles in Scotland since before the turn of the century, and when, in 1940, it became obvious that Scammell were unable to keep up with the demand for tank transporters and artillery tractors, Albion seemed a logical choice as a second supplier. Although the company had supplied quantities of both 3 and 10 ton cargo vehicles throughout WW2, we are concerned here only with the CX22S and FT15N/FT15NW artillery tractors, the CX24S tank transporter, and the CX33 experimental tank tractor.

The Albion WD-CX22S artillery tractor was produced to the same War Office specification as the Scammell 'Pioneer' R100 (see Chapter 5.1), and was based on the chassis and engine of the Company's commercial CX23N 10 ton cargo truck, which was also used by the War Office. Towards the end of the war, the Company also produced a limited number of experimental low-profile artillery tractors designated WD-FT15N and FT15NW: these did not see active service, but enjoyed a chequered career in the immediate post-war years with several suggestions being put forward as to possible uses to which they might be put.

The WD-CX24S 30 ton tank transporter was not successful and was never produced in large numbers, ending its days as a 20 ton tractor and semi-trailer outfit, used mostly for transporting engineering plant.

The final model, the eight-wheeled, twin-engine WD-CX33, was produced only in experimental form, as a tank tractor.

None of the vehicles described in this chapter was particularly successful but they do demonstrate the dire straits in which the War Office found itself for most of the war years, where almost any vehicle, no matter how poorly equipped for its intended role, was better than no vehicle at all.

After WW2 was over, Albion continued to supply military chassis for a number of years and was involved in the development of the post-war 'CT-rated' FV1300/FV14000 3/5 ton artillery tractor project.

CX22S ARTILLERY TRACTORS

During most of the war years, there was a shortage of heavy artillery tractors, and in 1943 Albion were approached to produce a 'sister' vehicle to the R100 Scammell 'Pioneer'. The Albion CX22S was based on

CHAPTER 2.1: ALBION CX and FT SERIES

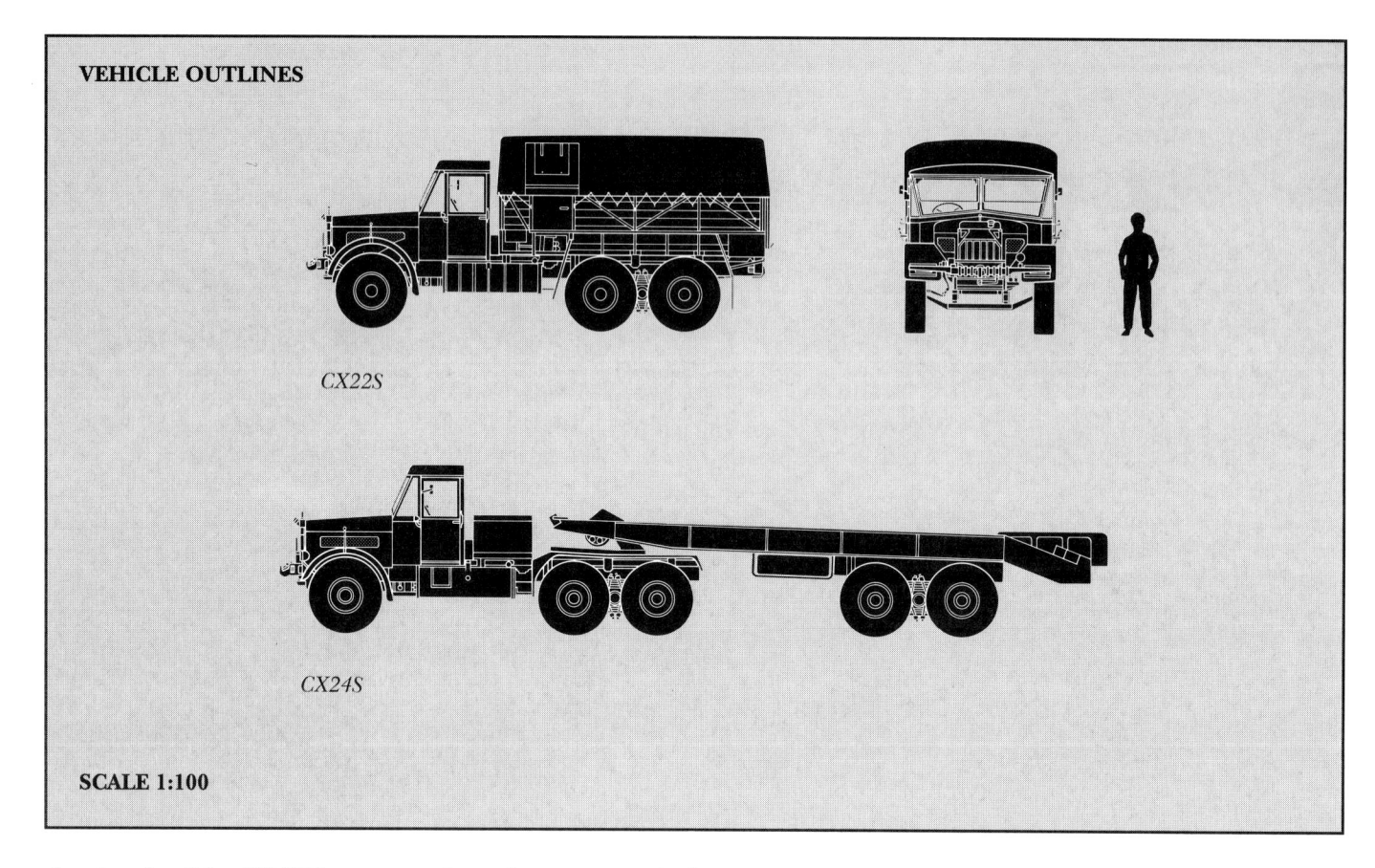

VEHICLE OUTLINES

CX22S

CX24S

SCALE 1:100

the chassis of the CX32N cargo truck, and was powered by an Albion EN244 six-cylinder diesel engine producing 100bhp. The vehicle was designed for towing the 7.2in and 6in howitzers and offered a comparable performance to that provided by the trusty Scammell.

The chassis incorporated a Scammell winch to facilitate manoeuvring the gun, but unlike the 'Pioneer' the body had no overhead gantry for loading, relying instead on skids. There was a three-man cab, and the body included accommodation for ten men. Like the CX24S tank transporter, the prototype employed a cab rather like that fitted to the 'Matador', but since the vehicle was of the normal control type, the resulting flat, upright windscreen and conventional bonnet gave a rather pig-like profile. During subsequent production, the windscreen was sharply raked, improving the frontal aspect. The cab was designed to be split at the waistline, and the body dismantled to make shipping easier.

The total production quantity of 532 vehicles was produced between November 1943 and June 1945, with the first vehicles being delivered in February 1944.

Although the CX22S remained in service after the war, once the pressures of actual fighting were out of the way, the War Office could obviously afford to be a bit more choosy and the Albion began to fall from favour.

In 1957, when the American-built Mack 6x6 tractors were being retired due to the difficulties of obtaining spares, both the Scammell and the Albion were proposed as possible replacements. In the end the Albion was 'considered unsuitable for deployment as a heavy artillery tractor', with three major reasons cited: firstly, there was no front wheel drive; secondly, the ground clearance (less than 325mm at the rear axle) was inadequate; and lastly, the rear tow hitch was too low. These had not apparently proved any kind of hindrance during WW2 but the priorities of the time were somewhat different.

However, somewhat surprisingly, the Albion CX22S was still quoted in the RAOC table of vehicle weights and measurements published in 1970, so presumably some remained in service well into the post-war years.

Nomenclature
Albion CX22S, tractor, 6x4, heavy artillery.

Specification
Dimensions and weight
Length: 7777mm.
Width: 2669mm.

Height: 3162mm.
Wheelbase: 4474mm ; rear bogie, 1372mm.
Track: front, 2083mm; rear, 2077mm.
Ground clearance: 241mm.

Weight: laden, 15,858kg; unladen, 10,627kg.

Performance
Speed: 44km/h maximum.
Fuel consumption: 0.35 litre/km.
Maximum range: 960km.

Turning circle: 18.91m.

Capacity: 5275kg.

Description
Engine
The engine fitted to the CX22S was a six-cylinder Albion diesel unit, type EN244. The engine was a monobloc casting, with press-fitted dry cylinder liners; the overhead valves were operated by short pushrods, driven by a high-mounted camshaft. Lubrication was by conventional wet sump.

Capacity: 9060cc.
Bore and stroke: 4.625 x 5.5in.

Power output: 100bhp at 1750rpm.
Maximum torque: 366 lbs/ft at 1000rpm.

Transmission
The engine was arranged to drive through a dry-plate clutch to a four-speed and reverse gearbox, with a two-speed transfer case. The drive was transmitted via the transfer gears through conventional propeller shafts to the rear axles.

A small compressor, intended to provide compressed air for the braking system, was mounted on the gearbox.

Suspension and axles
The front axle was an 'I' section beam-type unit, mounted on semi-elliptical springs. The rear axle bogie was suspended on inverted semi-elliptical springs, with two springs either side, mounted one above the other, and pivoting on central trunnion mountings. Torque control was by means of a pair of bushed arms on each bogie. The axle drive was by means of overhead worm gear.

Steering gear
Steering was by conventional worm-and-sector steering box, with a short drag link connecting the drop arm to the right-hand hub, and with a track rod running from one hub to the other. There was no power assistance.

Braking system
Braking effort was provided only on the rear wheels, with both the service and parking brakes being operated mechanically, with air-pressure servo assistance. Twin-line air connections were provided for the trailer brakes.

Road wheels
Wheels were 20in diameter, mounting standard 14.00x20 tyres. The rear wheels were shod with cross-country type tyres, while the fronts, which were undriven, were fitted with standard road-pattern tyres.

Chassis
The chassis was a conventional ladder frame design of pressed steel, running straight from front to rear, with channel section main and cross members.

Cab and bodywork
On the prototype, the cab was originally similar to that fitted to the AEC 'Matador' although obviously with the addition of a separate engine compartment. Production vehicles used a composite wood-and-steel three-man cab, with a raked windscreen, front quarterlights, and narrower doors; the canvas-covered roof had a shallower curve, and there was a combined roof lookout/anti-aircraft hatch fitted above the passengers' seat. The cab could be easily split at the waistline to reduce the shipping height.

The well-pattern rear body provided bench seating for four personnel, together with an additional two folding seats at the extreme rear; there was also stowage for ammunition and other supplies. The body was of composite wood-and-steel construction, and was provided with detachable hoops supporting a canvas cover; the top section of the cover was of metal on the prototype, rather in the style of the early 'Matadors'. Crew access doors were provided on either side and there was an observation position in the canvas cover. Ammunition loading skids were carried on the outside of the body.

Winch
An 8 ton Scammell vertical-spindle winch was fitted beneath the rear body, and was provided with 131m of steel cable. Fairlead rollers were provided at both the front and rear.

Electrical equipment
A hybrid 12V/24V system was employed: series/parallel connections were provided 24V for starting, and 12V for the lighting and auxiliary circuits.

CX24S TANK TRANSPORTER

At the end of 1940, there remained a serious shortage of heavy tank transporters, and with a maximum output of 17 vehicles a month, production capacity at Scammell's Watford plant was not sufficient to make-up the shortfall. In an effort to overcome this, the Ministry of Supply placed the first of three orders with Albion Motors for

CHAPTER 2.1: ALBION CX and FT SERIES

some 800 examples of what was a virtually unproven design of 30 ton tractor and purpose-designed semi-trailer - the CX24S.

The vehicle, which was of 6x4 configuration, was powered by a six-cylinder Albion EN248B petrol engine producing 140bhp, at that time the largest standard commercial engine produced in the UK. The tractor itself was based on the CX22S heavy gun tractor, and was adapted to provide an articulated configuration. The vehicle was primarily intended to carry the Stuart M3 'Honey' light tank on paved highways.

The non-detachable trailer, which was unique to this particular tractor, was a twin-axle design, with eight wheels in four pairs, mounted on a rear bogie similar to that used on the tractor. The carrying bed was almost parallel to the road, and folding, triangulated ramps were provided at the rear to facilitate loading; there was also a Scammell 8 ton winch to provide assistance. The first examples were fitted with a very old-fashioned looking cab where the windscreen was virtually upright, not unlike that fitted to the AEC 'Matador', but with a normal bonnetted engine compartment; these were soon replaced by a similar cab to that used on the CX22S artillery tractor.

Deliveries began in 1942, with some 350 examples despatched almost immediately to North Africa, with the remaining 450 or so employed in the UK.

Unfortunately, the vehicle was seriously under-powered and, since it was essentially a road-going design, was unable to cope with the rugged conditions it encountered. To add insult to injury, the engine was also subject to all-too frequent failure, with crankshafts prone to breakage due to over-revving when operators attempted to use the gearbox to supplement the inadequate brakes during long descents.

It was to no avail that Albion protested that the vehicle was being put to uses for which it had not been designed, but the truth was the CX24S was really only capable of carrying the lightest of the then-current tanks... and then only on metalled roads. Albion struggled to remedy the defects, largely at their own expense, but it was not long before the vehicle was down-rated to 15 tons capacity. Although a few examples soldiered on in the role for which they had been designed, the majority was allocated to other haulage tasks, most notably the transport of cable drums.

Nomenclature
Albion, CX24, 20 ton, semi-trailer, tank transporter, 6x4-8; later downgraded to transporter, 15 ton.

CX22S artillery tractor (TMB)

Same vehicle from rear (TMB)

Post-war - a very tidy CX22S tractor awaits disposal (KP)

CHAPTER 2.1: ALBION CX and FT SERIES

CX24S tank transporter and trailer (TMB)

Also from the same series, presumably taken at the Albion works (TMB)

Prototype CX33 tank tractor (TMB)

Specification

Dimensions and weight
Length: tractor only, 6710mm; semi-trailer only, 8388mm; tractor and semi-trailer complete, 13,355mm.
Width: tractor only, 2388mm; semi-trailer only, 2870mm.
Height: to top of cab, 2540mm; semi-trailer with ramps raised, 1372mm; semi-trailer to front stop, 1880mm.
Wheelbase: 4498mm; rear bogie, 1372mm; semi-trailer bogie, 1372mm.
Ground clearance: 279mm.
Track: front, 2028mm; rear, 1910mm.

Weight: laden (carrying a 'Crusader' tank), 34,898kg; unladen, 15,413kg.

Performance
Speed: 30km/h.
Fuel consumption: 0.97 litre/km.
Maximum range: 470km.

Turning circle: right lock, 21.057m; left lock, 21.082m.
Maximum gradient: early models, 32%; later models (with different transfer gears), 24%.

Capacity: 19,485kg.

Description

Engine
The CX24S was fitted with a six-cylinder Albion petrol engine, type EN248B. The engine was a monobloc casting, with press-fitted dry cylinder liners; the overhead valves were operated by short pushrods, driven by a high-mounted camshaft. Lubrication was by conventional wet sump. A type RZVGO Solex carburettor was fitted.

Capacity: 10,487cc.
Bore and stroke: 4.75 x 6in.

Power output: 146bhp at 2100rpm; later downgraded to 1850rpm.
Maximum torque: 462 lbs/ft at 900rpm.

Transmission
Drive was transmitted through a single dry-plate clutch to a four-speed and reverse gearbox, with a two-speed transfer case. The transfer ratios were reduced on the later models in an effort to improve performance. The drive was transmitted via the transfer gears through conventional propeller shafts to the rear axles.

A small gearbox-mounted compressor provided air for the braking system.

Suspension and axles
The front axle was an 'I' section beam, mounted on semi-elliptical springs. The rear axle bogie was suspended on a pair of inverted semi-elliptical springs either side, arranged to pivot on central trunnion mountings. Torque control arms were attached to the axles and chassis.

CHAPTER 2.1: ALBION CX and FT SERIES

Drive to the wheels was by means of overhead worm gear and fully-floating drive shafts.

The trailer bogie was very similar to that fitted to the tractor.

Steering gear
The steering box incorporated worm-and-sector gears, with a short drag link connecting the drop arm to the right-hand hub, and a track rod running from one hub to the other. There was no power assistance.

Braking system
Braking was provided on all four tractor wheels, with both the service and parking brakes being operated mechanically with air-pressure servo assistance; should the air system fail, braking effort remained available at the rear tractor wheels by means of the mechanical linkages. Twin-line air connections were provided for the trailer brakes.

The trailer parking brakes were applied by handwheel, through a compensating bar.

Road wheels
The steel wheels were 20in diameter, mounting standard road-pattern 14.00x20 tyres; both the tractor and trailer bogies were fitted with twin wheels.

Chassis
The chassis was a conventional ladder frame design of pressed steel, running straight from front to rear, with channel section main and cross members.

Semi-trailer
The trailer was of steel channel and platform design, fitted with steel runways parallel to the road. The runways were fitted with outer guide rails and there were adjustable stop blocks at the trailer nose; folding triangulated ramps were provided at the rear for loading. A spare wheel was carried below the trailer bed.

The fifth-wheel was non-detachable, and was fitted immediately over the centre of the rear tractor bogie. There were no support legs for the trailer.

Cab and bodywork
Like the CX22S model, the prototype cab was similar to that used on the AEC 'Matador' giving a very 'perpendicular' appearance.

Production vehicles used a composite wood-and-steel three-man cab, with a raked windscreen, front quarterlights, and narrower doors; the canvas-covered roof had a shallow curve, and there was a combined roof lookout/anti-aircraft hatch fitted above the passengers' seat. The cab could be easily split at the waistline to reduce the shipping height.

Low-silhouette FT15N artillery tractor (TMB)

Same vehicle from rear (TMB)

Looking slightly battered, this FT15N differs in many details (IWM)

CHAPTER 2.1: ALBION CX and FT SERIES

Similarly, from the rear there are many detail differences (IWM)

Was this a prototype FT15N cargo truck? (IWM)

Prototype for FV14000 - was this based on the FT15N? (TMB)

On the prototype, the winch (see below), which was installed behind the cab, was exposed, while on production vehicles, a box-type enclosure was fitted to provide weather protection.

Winch

An 8 ton Scammell vertical-spindle winch was fitted behind the cab, provided with 188m of steel cable; a torque limiting device was provided to protect the winch from overload. Fairlead rollers were provided at both the front and rear.

Electrical equipment

A series/parallel connected 12V/24V electrical system was employed, providing 24V for starting, and 12V for the lighting and auxiliary circuits.

FT15N/FT15NW ARTILLERY TRACTORS

In 1945 and 1946, Albion produced about 150 examples of a low-profile experimental medium field artillery tractor with a semi-forward control cab, and 6x6, 5 ton chassis configuration. Powered by a six-cylinder Albion petrol engine, the model was designated FT15N, and FT15NW in fully-waterproofed form. Unfortunately, although the vehicle possessed a reasonable performance, it probably arrived too late to make any real contribution to the war effort and various attempts were made to find a role for this unusual beast in the post-war years.

During 1951, when the 3 ton CT Vauxhall FV1300 project was falling apart through cost increases and development delays, Albion proposed that the FT15N chassis be put back into production with the Rolls-Royce B80 engine, to provide an interim 'combat' type tractor. The projected cost was approximately £3000 per vehicle against an estimated price of £3700 (at that time) for the Vauxhall FV1300. Albion had stated that all of the tooling still existed and deliveries could begin within 52 weeks of placing the order, with some 2000 vehicles being made available over a two-year period. Two of these vehicles had already been fitted with B80 engines and used by FVRDE since 1946 as a mobile test bed with apparently-promising results, and it was suggested that by uprating the springs, the capacity could be increased to 5 tons. This was an odd statement to make since the chassis was supposedly rated at 5 tons already: perhaps a post-war ton was heavier than a WW2 ton!

However, nothing was to come of the proposal in this form, although Albion did take over the hapless FV1300 project with which Vauxhall were having such difficulties. It is quite possible that the prototypes which Albion did produce for this project, by that time designated FV14000, were actually based on the FT15N chassis. The design was of a similar semi-forward control configuration, with the same 3810mm wheelbase and 1372mm bogie centres. Certainly the Albion vehicles were nowhere near as complex as those which Vauxhall had produced, and by the time Albion were involved, the weight rating had been increased from 3 tons to 5 tons, and the vehicle downgraded from CT to GS.

Anyway, whatever the truth of the matter, the FV1300/FV14000 was never to make it into production in any form, whether based on the FT15N or not.

However, the FT15N was not dead yet.

In 1953, it seemed that although virtually everyone at the War Office had finally been forced to admit that the FV1300/FV14000 project was going extremely badly, the final decision to abandon it had not been taken. At the time, there was a suggestion that these vehicles, which were quoted as having 'a good though slow cross-country performance' might be used to provide a stopgap as a tank supply vehicle in place of the FV1300, until deliveries of the new equipment began.

The offer was not accepted and the decision was taken to struggle on with the FV1300/FV14000 Series though apparently no final decision had been taken on whether the vehicle was to be rated at 3 tons or 5 tons.

There seemed to be no further uses for the FT15N, but at least one of the tractors was demobbed and survived into 'civvy street' equipped with a fixed jib in place of the artillery tractor rear body, and used as a commercial recovery tractor. Another remains in preservation in the North East, but the remainder simply faded away, unwanted and unloved... they were probably cut-up for scrap.

Nomenclature and variations
Albion FT15N/FT15NW, tractor, 6x6, field artillery.

The FT15NW model was waterproofed, but was otherwise identical.

Although the vehicle does not appear in any contemporary documentation, there is photographic evidence to suggest that there was also a low-profile cargo equivalent of the FT15N artillery tractor. The cargo vehicle appears to be based on the same chassis and cab, but the longer engine compartment suggests that perhaps a different power unit was fitted. The photographs which exist were taken at FVRDE and possibly are of just a single experimental vehicle.

Specification
Dimensions and weight
Length: 6375mm.
Width: 2324mm.
Height: 2337mm.
Wheelbase: 3810mm; rear bogie, 1372mm.

Weight: unladen, 6007kg.

Performance
No figures available.

Description
Engine
The engine fitted to the FT15N was a six-cylinder Albion side-valve petrol-fired unit, designated EN281. Lubrication was by conventional wet sump.

Capacity: 4566cc.
Power output: 95bhp.

Transmission
Drive was transmitted through a single dry-plate clutch to a four-speed and reverse gearbox, with a two-speed transfer case, then by conventional propeller shafts to the axles.

Suspension and axles
The axles were suspended on semi-elliptical springs, with two springs at the front, and four inverted springs at the rear arranged to pivot on central trunnion mountings. The rear bogie was similar to that fitted to other tractors in the CX series, with torque control arms attached to the axles and chassis.

Drive to the wheels was by means of overhead worm gear and fully-floating drive shafts.

Steering gear
The steering box incorporated worm-and-sector gears, with a short drag link connecting the drop arm to the right-hand hub, and a track rod running from one hub to the other. There was no power assistance.

Braking system
Braking was hydraulic with air-pressure servo assistance. The parking brake was mechanical, operating on the rear wheels only.

Road wheels
Wheels were 20in diameter, mounting cross-country pattern 10.50x20 tyres.

Chassis
The chassis was a conventional ladder frame design of pressed steel, running straight from front to rear, with channel section main and cross members.

Cab and bodywork
The cab was a semi-forward control type, with steel wings, roof and engine compartment, and canvas doors. There was a combined roof lookout/anti-aircraft hatch fitted above the passengers' seat.

The low-profile, well-pattern rear body provided bench seating for the gun crew, together with stowage for ammunition and other supplies. The body was of composite wood-and-steel construction, and was provided with detachable hoops supporting a canvas cover; canvas crew access doors were provided on either side.

Winch
An 8 ton worm-drive Turner winch was fitted behind the cab, with fairlead rollers provided for both front and rear winching.

Electrical equipment
A series/parallel connected 12V/24V electrical system was employed, providing 24V for starting, and 12V for the lighting and auxiliary circuits.

CX33 TANK TRACTOR

Albion's final foray into military heavy haulage was the unconventional CX33 ballast tractor tank transporter, supposedly designed as an investment against a possible shortage of the American-built Diamond T's. As something of a portent of the FVRDE post-war penchant for mechanical complexity, the CX33 was as complicated as the rugged Diamond T was simple.

The low-profile, eight-wheeled CX33 was arranged in an unusual close-coupled 8x8 configuration, broadly following the style of the similar 6x6 vehicle produced by Dennis Brothers in 1943. The Dennis had originally been conceived as an 8x8 artillery tractor, and was equipped with a pair of Bedford engines. It differed from the Albion in having no suspension, using balloon tyres for springing in the style of modern earth-moving equipment, and used a skid-steering system. The name 'Octolat' with which the Dennis was saddled was presumably derived from its original eight-wheeled configuration, and the acronym LAT for light artillery tractor.

The Albion had every appearance of being a product of the same school of design. Also originally produced in 1943, the Albion had a maximum straight-line pull of 75 tons, was designed for use both as a tank-transporter tractor, and for removing disabled tanks from the battlefield.

A prototype was produced using a pair of six-cylinder petrol engines, similar to those installed in the CX24S chassis, but this time designated EN248E, and producing 140bhp each. The two engines were mounted side-by-side in the chassis, each with its own clutch and gearbox. The first and fourth axles were driven by one engine, the second and third axles were driven by the other, though in 1945, a later version of the prototype, or possibly a second vehicle, was modified to an 8x6 configuration with the second axle undriven. The 1943 prototype was steered by the outer axles, while the later version used the front two axles for steering.

A cab was installed at either end of the tractor; the front cab, obviously being allocated to the driver, with the rear cab intended for the winch operator. Both cabs, which of course did not require an engine compartment, appear to have been based on the cab used for the CX22S and CX24S models. A ballast box was installed behind the engines and between the cabs.

Copies of drawings held by the Tank Museum, unfortunately undated, show alternative approaches to the design. One placed the driver at the front in a low, streamlined cab, with the engines and transmission at the rear. The drawings for this version did not include a winch, although there was plenty of space, and no detail was shown on the bodywork between the cab and the engines, so it could well have been sited in that position. Another version showed the engines placed behind the driver, with the gearboxes and clutch housings forward of the engine, beneath the driver, and a horizontal spindle winch immediately behind the engines. The final version, which most closely matched the prototype, placed the winch in an open position at the rear; in this version, the winch appeared to be a vertical-spindle Scammell-type machine.

Only one or two prototypes seem to have been constructed, and neither was adopted by the War Office. Both quietly vanished without trace at the end of the war.

Nomenclature
Albion CX33, tractor, heavy, 8x6, tank transporter.

Specification
Dimensions and weight
Length: 7720mm.
Width: 2680mm.
Height: 2710mm.
Wheelbase: 4116mm; bogie centres, 1372mm.

No other figures are available.

Description
Engine
The CX33 was also equipped with an EN248 six-cylinder Albion petrol engine, like the CX24S chassis, but this time designated type EN248E.

The engine was a monobloc casting, with press-fitted dry cylinder liners; the overhead valves were operated by short pushrods, driven by a high-mounted camshaft. A type RZVGO Solex carburettor was fitted, and lubrication was by means of a conventional wet sump. Two cooling radiators were provided, one for each engine, positioned in the centre of the chassis.

Capacity: 10,487cc.
Bore and stroke: 4.75 x 6in.

Power output: 146bhp at 2100rpm; later downgraded to 1850rpm.
Maximum torque: 462 lbs/ft at 900rpm.

Transmission
Each engine was provided with a separate dry-plate clutch and gearbox, in unit construction with the engine.

The gearboxes were positioned at the front of the tractor, with the side-mounted linkages connected together to a single lever. The drive was transmitted by means of transfer gears, probably with a choice of two ratios, via conventional propeller shafts to the axles. One engine was connected to the first and fourth axles, while the remaining axles were driven by the second engine. The drive-line arrangements were subsequently modified to 8x6, with the second axle undriven.

A small compressor was mounted on each gearbox, probably driven directly by the layshaft.

Suspension and axles
The four axles were arranged as two bogies, suspended on possibly two pairs of inverted semi-elliptical springs, mounted one above the other. Torque control was by means of a pair of bushed arms on each bogie.

The wheel centres (1372mm) and general design features suggest that these might well have been the same bogies used on other vehicles in the Albion CX range.

Steering gear
The vehicle was of forward-control configuration, with a centrally-positioned steering wheel connected to a conventional steering box. The 1943 prototype was steered by the outer axles, with the wheels on each axle arranged to turn in the opposite direction. The later version used the front two axles for steering.

Braking system
The Tank Museum drawings suggest that braking was provided only on the two centre axles. The drawings clearly show braking system air lines and reservoirs, and if the braking system were based on that used on the other vehicles in the CX series, operation of both the service and parking brakes would have been mechanical with air-pressure servo assistance. Twin-line air connections would have been provided for the trailer brakes.

Road wheels
Wheels were 20in diameter, mounting standard 13.50 or 14.00x20 cross-country type tyres.

Chassis
The channel section steel chassis was a conventional ladder frame design, running straight from front to rear, with mountings for the engine, transmission, and axles.

Cab and bodywork
The bodywork on the prototype consisted of two similar cabs, clearly based on that used on the CX22S artillery tractor. One cab was placed at either end, with a steel ballast box positioned amidships, and a square engine compartment.

Winch
The winch, which appeared to be a Scammell 15 ton vertical-spindle unit, similar to that used on the CX22S artillery tractor, was installed at the rear of the chassis with the controls fitted in a separate cab.

DIAMOND T
CHAPTER 3

CHAPTER 3: DIAMOND T

There's something about American names, which to European ears, lends them a touch of magic. It hardly matters whether they are naming perfectly-commonplace automobiles, like Cadillac and Pontiac, or whether, like Albuquerque or Baton Rouge they are the names of unexciting, provincial towns. Somehow Reliant and Barking just don't have the same ring.

Diamond T is definitely one of those names. Even though it came about for the most prosaic of reasons, it seems to conjure up all kinds of romantic images.

In 1905, the Diamond T Motor Car Company was founded by one Charles Arthur Tilt, son of a footwear manufacturer who had called his top-of-the-line shoe range Diamond T. Following the old adage 'like father, like son', when he started making motor cars from a Chicago factory, Charles adopted the same name and logo that his father had used - the 'Diamond' stood for quality, the 'T' for Tilt.

For the first six years the company built quality motor cars, but in 1911, Diamond T switched over to truck production, never looking back.

Although they were never to build their own engines or running gear, this was a common enough arrangement in the USA, and soon Diamond T had acquired a reputation for producing trucks which were both tough and good looking. It was not long before the military started to take notice, and during the Great War, Diamond T was one of 15 manufacturers of the standardised 'Liberty' truck, producing some 638 examples. In 1930, the US Army ordered a quantity of 2.5 ton 6x6 anti-aircraft vehicles based on a Diamond T chassis, and in 1934, took delivery of a number of 1.5 ton 4x2 reconnaissance trucks, both chassis being powered by a proprietary six-cylinder Hercules diesel engine.

The company continued to supply medium and heavy trucks to both the commercial and military markets throughout the 'thirties', but as for so many automotive manufacturers, WW2 brought genuine prosperity, with the Chicago factory often working 24 hours a day to satisfy the military orders.

By the time the British Purchasing Commission had placed their first orders for the model 980 ballast tractor in 1941, Diamond T was already busy supplying 4 ton chassis to the Canadian, and ultimately to the US Army. In all, some 50,000 chassis were supplied to the Allies during the years 1940-45. Of these, more than 12,500 were the 4 ton 6x6 model 975/975A, bodied as cargo vehicles, machinery trucks, bridging vehicles, and mobile cranes. There were also quantities of the model 969 6x6 wrecker, and of course the model 980/981 ballast tractors which were to remain in service with the British for close on 30 years. Similar chassis were also produced for the US commercial market throughout the war years.

Although the Hercules JXFE diesel engine was the standard offering for the military chassis, Hall-Scott petrol engines were also used on commercial vehicles, and the British 980/981 tractors were upgraded with Rolls-Royce C6 diesels in the 'fifties.

The 4 ton chassis were adopted as 'standard' by the US Army and remained in service into the 'fifties. The ballast tractor was rated 'substitute standard' but later downgraded to 'limited standard' and, despite its valiant service with the British, was never fully embraced by the US Army.

In 1947, Diamond T produced their first post-war design, surprisingly, equipped with a petrol engine. But the military had frozen all orders by that time, and although the commercial market was not really strong enough to produce sufficient sales, the company managed to keep its head above water for another 20 years. At one time there were even sufficient European sales to warrant an assembly plant in the Netherlands. However, increased competition from the domestic producers, and the ever-rising price of fuel, meant that the last European Diamond T was assembled in 1954.

The Korean war and the threat of East-West conflict in the early 1950's provided a small boost for military production, and Diamond T constructed examples of the standardised M54 6x6 5 ton cargo truck, and the M62 6x6 5 ton wrecker for the US Army.

The American White company had already taken over the old-established Freightliner, Sterling, and Autocar names, when in 1957, they also acquired Reo. In 1967, unable to keep the company afloat as an independent any longer, Diamond T joined forces with Reo, and thus became part of the White empire. The decision was taken to merge the Diamond T and Reo names and for a while the trucks were badged as Diamond-Reo, retaining elements of both companies' logos. But the writing was already on the wall, and in 1974, use of the Diamond-Reo name was discontinued and another motoring pioneer disappeared for ever.

However, the Diamond T is one of those trucks which promotes loyalty and affection beyond its level of commercial success, and fortunately many 'Dizzy T's' remain in the hands of collectors across the globe.

CHAPTER 3.1

DIAMOND T 980/981
M20

While the Pacific M25 and M26 'Dragon Wagons' may have been the largest of the WW2 trucks, they were both ugly beasts, and anyway, apart from some post-war experiments, were really only used by the Americans. The best-looking of the big trucks was indisputably the 12 ton Diamond T prime mover. Not only was the Diamond T initially built specially for the British, but it remained in service for almost 30 years after the war was over, and even retains its good looks to this day.

The American-built M20 Diamond T prime mover had been produced under the direction of the US Quartermaster Corps to satisfy a British requirement for a ballast tractor suitable for use in the heavy tank transporter and tank recovery roles, to supplement the British-designed 30 ton Scammell 'Pioneer'. The tractor was purchased by the Ministry of Supply in quantity even before the introduction of the 'Lend-lease' arrangements, and was first used by the British Army in North Africa, where it must have made the enormously capable but rather ungainly 'Pioneer' look positively ancient.

Originally designed for use with the American Rogers M9 45 ton trailer, but also used with similar trailers produced by Cranes, Dyson, Fruehauf and others, the complete outfit was described as 'tank transporter M19' and was rated at 40 tons. However, in an attempt to overcome the shortage of off-road tank transporters, it was also modified towards the end of the war by fitting a turntable to handle the Shelvoke & Drury semi-trailer normally associated with the 30 ton 'Pioneer' tractors.

The Diamond T was both powerful and capable, but the lack of front-wheel drive meant that its off-road performance was somewhat restricted, and the outfit was relatively easily defeated by a build-up of soft soil in front of the trailer wheels. Admittedly the Scammell had no front wheel drive either but its balloon tyres were more effective at pushing through difficult terrain.

Available in both closed and open cab configurations, there were two variants, one with a Gar Wood 300ft winch pulling to the rear only, the other with a 500ft winch modified to provide front pulls as well; these were designated models 980 and 981 respectively.

The 'Dizzy T', as it came to be known, also served with the US Army during WW2, and many passed into civilian service right across Europe in the post-war years, often forming the mainstay of commercial heavy haulage fleets. Those vehicles which remained with the British

TUGS OF WAR

CHAPTER 3.1: DIAMOND T 980/981

VEHICLE OUTLINES

Closed-cab 980/981 ballast tractor

Open-cab 980/981 ballast tractor

Closed-cab 980/981 fifth-wheel tractor with Shelvoke & Drury semi-trailer

SCALE 1:100

SPECIFICATION

Dimensions and weight

	980	981	980-C6N	981-C6N
Dimensions (mm):				
Length	7092	7218	7092	7218
with semi-trailer	15,189	15,189	-	-
Width	2540	2515	2540	2515
with semi-trailer	2946	2946	-	-
Height:				
overall	2484	2484	2484	2484
to top of closed cab	2477	2477	2477	2477
to top of open cab	2591	2591	2591	2591
Wheelbase	4553	4553	4553	4553
rear bogies	1321	1321	1321	1321
Track:				
front	1927	1927	1927	1927
rear	1880	1880	1880	1880
Ground clearance	454	454	454	454
Weight (kg):				
Gross laden weight*	19,270	19,270	23,266	23,266
with semi-trailer**	54,371	54,371	-	-
Unladen weight***	12,142	12,142	11,480	11,480
with semi-trailer	22,858	22,858	-	-
Bridge classification:				
solo	20	20	24	24

Performance

	980	981	980-C6N	981-C6N
Maximum speed (km/h)	37	37	47	47
Fuel consumption (litre/km)	1.30	1.30	0.81	0.81
Maximum range (km)	440	440	676	676
Turning circle (m)	24.69	24.69	24.69	24.69
Approach angle	45°	45°	45°	45°
Departure angle	51°	51°	51°	51°
Maximum gradient	14.75%	14.75%	12.5%	12.5%
Fording, unprepared (mm)	813	813	559	559
Capacity:				
Maximum towed load	52,272	52,272	69,444	69,444

* Laden weight depends on ballast carried; figures quoted include normal maximum of 6650kg for 980/981 models, and 11,250kg for 980-C6N/981C6N.
** Laden with Sherman tank.
*** The open cab resulted in a weight saving of 136kg.

were up-engined with the Rolls-Royce C6 diesel during the 'fifties and many were still in service until well into the 'seventies.

DEVELOPMENT

Rated at just 30 tons, the Scammell 'Pioneer' tractors (see Chapter 5.1) were insufficiently powerful to haul the near-40,000kg weight of the Churchill tanks, and anyway were also in short supply for most of the war. Aside from Scammell, there was no other British truck company either with the expertise or factory capacity required to produce a heavier prime mover, and in 1940 the Washington-based British Purchasing Commission was asked to approach the US Quartermaster Corps with a view to involving an American company in developing and producing a suitable vehicle.

Approaches were made to a number of companies, including Mack, Ward La France, Diamond T, and White. Eventually, contractual arrangements were made with the Diamond T company of Chicago, who had a reputation for building tough, good-looking trucks which were well-suited to the rugged road conditions existing in the USA at that time, and who were already supplying trucks to the Canadian Army.

Design work began in 1940. The chassis was specially produced by Diamond T for this application, while the cab was borrowed from the contemporary commercial 4 ton chassis; the axles were supplied by Timken, and the transmission produced by Fuller. The British wanted a diesel engine, and so the Hercules DXFE power plant was selected, producing 201bhp from a 14.5 litre capacity. The engine, which was one of a series of 21 heavy-duty models originally produced in 1931 and including two-, four-, six- and eight-cylinder variants, was manufactured by the Hercules Motors Corporation of Canton, Ohio.

The resulting truck was known as the Diamond T model 980, and there was subsequently a model 981. With its distinctive long 'coffin' nose and short ballast body, the truck became a favourite of the tank transporter crews, and was quick to prove itself in action.

There were obvious differences in design philosophy between the US-designed truck and the British Scammells. For example, unlike the counterpart Scammell, which employed a coupled semi-trailer, the Diamond T was a ballast tractor. This gave far greater flexibility in use, particularly when employed for tank recovery rather than simple transportation, for the tractor could leave its trailer in position and be manoeuvred around a disabled AFV to allow easier loading using the tractor winch.

The Diamond T employed twin rear wheels, which the British maintained were more likely to trap mud and stones and lead to punctures than the larger single rear wheels used on the Scammell. And of course, all of the Diamond T's were left-hand drive.

In 1941, the British ordered a total of 485 Diamond T tractors, together with 285 of the 45 ton Rogers trailer, and some 200 similar trailers, albeit rated at 40 tons, from British manufacturers. The first trucks were delivered in April 1942, when a total of some 130 tractors and trailers were used in the Tunisian campaign.

In 1943, in line with the general trend for reducing the unnecessary consumption of materials, an open-cabbed variant was produced. This saved steel, but at the same time, with the top and windscreen folded, also reduced the overall height of the vehicle.

Despite its rugged appearance, the Diamond T was not without problems, the most serious of which could lead to a complete loss of the rear bogie. Other difficulties included cracked cylinder blocks, valve seat failures, and breakage of the injector shafts; the engine was also prone to rod breakage following over-revving.

Unlike the British-designed winches, the Gar Wood machine fitted to the Diamond T was not equipped with paying-on gear and this could lead to premature damage and breakage of the winch cable due to kinking and jamming during operation. This problem was overcome in mid-1944 by the retrospective fitting of a British-designed pendulum paying-on device.

Other modifications included the raising of the ballast body by 225mm to provide clearance for overall tracks which were sometimes fitted to the rear wheels after the style of the Scammell 'Pioneer', and the provision of a fifth-wheel turntable which allowed any Diamond T ballast tractor to be converted to tow a semi-trailer.

The Diamond T was a tough and powerful truck. If the Hercules engine proved insufficient for a particularly tricky task, the vehicles could be double-headed to provide massive pulling power, and the Diamond T was one of those trucks which commanded respect from both the British and American troops.

Modifications

The model 980/981 continued more-or-less unchanged throughout the war years, and while the Americans probably replaced the model soon after the war was over, more than 1200 vehicles remained in service with the British army, many for a further 30 years, getting uprated from 40 tons to 50 tons at some time along the way. But as time marched on it must have become obvious that if the vehicle was to continue to give useful service, some

CHAPTER 3.1: DIAMOND T 980/981

upgrading of its performance was required; its top speed, for example was just 34km/h and a considerable degree of skill with the gearbox was required to get the best out of the engine which produced its peak power at exactly the moment when a gear change was called for.

In 1952/53, a Hercules-engined Diamond T was experimentally fitted with a supercharger at the Leyland factory. Air was drawn through a pair of oil-bath air cleaners mounted above the cab roof, with steel trunking connecting them to a twin-vane Roots-type blower mounted on a specially-cast manifold. At more-or-less the same time, Rolls-Royce, who were well into the manufacture of the standardised 'B Series' petrol engines at the time, fitted at least one Diamond T tractor with a six-cylinder 'C Series' diesel engine.

These experiments may well have been an attempt to provide comparable performance to that available from the then-new Thornycroft 'Antars' or may simply have been an exercise in modernising the ageing Diamond T's. In the event, although the supercharger experiment was not pursued, of course the American Hercules engines were not getting any younger and, apart from the standardisation problem, parts were probably not getting either cheaper or easier to come by.

The Rolls-Royce experiment was to prove to be the way forward. In 1953 when the Director Weapons Development (DWD) visited the Rolls-Royce factory at Crewe, among other topics, he was to emphasise the importance of the outcome of the engine replacement programme. In the words of Lt Col W P St John Becher of the War Office General Staff, writing in March 1953, this would 'save... the Diamond T repair programme'. Apparently, at that time, some 700 of the 1000 British tank transporter tractors were the trusty Diamond T.

Although the Rolls-Royce engine was actually no more powerful than the Hercules which it replaced, the experiment proved to be a success, and between 1956 and 1957, most of the remaining Diamond T tractors were equipped with the Rolls-Royce C6 diesel engine by REME workshops. The Rolls-Royce engine produced its power over a broader rev band, and this brought about a useful increase in both acceleration and top speed, as well as providing an extended range. Of course, it also brought the Diamond T in line with other heavy vehicles on the inventory at the time, but it seems that not all of the vehicles were converted. In the 1970 edition of the RAOC table of weights and dimensions for 'A', 'B' and 'C, vehicles, both Hercules and Rolls-Royce engined Diamond T tractors were still listed.

There was also some confusion over the issue of the documentation for the modified tractors. A new edition

Closed-cab 980 ballast tractor (TMB)

Rear view of 981 ballast tractor (TMB)

Closed-cab 980 30 ton tractor with Scammell semi-trailer (REME)

CHAPTER 3.1: DIAMOND T 980/981

Open-cab 981 ballast tractor with 40 ton Mk1/Mk 2 trailer (TMB)

30 ton tractor with Shelvoke & Drury semi-trailer (TMB)

Post-war 981 open-cab ballast tractor with Rolls-Royce C6 diesel (KP)

of the User Handbook covering the Rolls-Royce engined 980 and 981 Diamond T chassis was issued in July 1958, while the EMER detailing the work, EMER S532/1 Supplement 1, was not published until October of the following year. This would suggest that, following the trials, perhaps some pilot models were initially produced to test the production feasibility of the project, and were issued for unit trials, and that once this was deemed satisfactory, the bulk of the vehicles would be converted at REME base workshops. There was a considerable amount of work involved, and the EMER described new front and rear engine mountings, modifications to the chassis, bonnet and wheel arches, new cab fittings, new clutch, new braking system compressor, and additional electrical equipment.

However, the conversion gave the venerable Diamond T a new lease of life, and a number remained in service until 1975 when they were presumably finally phased out in favour of the Mk 3/3A 'Antars' (see Chapter 6.1).

Production

When the USA entered the war in 1942 the Diamond T was also employed as a tank transporter and recovery vehicle by US Army units.

Despite its performance, the Americans were obviously never completely happy with what was essentially a compromise design and the Diamond T was never fully standardised. It was originally designated 'Substitute Standard', but in June 1943, with the adoption of the M25 and M26 Pacific tractors, the rating was downgraded to 'Limited Standard'. However, there was a continuing shortage of vehicles throughout the war because so many of the Diamond T production was allocated to the British.

A total of 5871 Diamond T 980 and 981 tractors was constructed between 1941 and 1945, including a number constructed under subcontract by FWD. An initial quantity of 200 tractors was diverted to the US Army, and there may well have been subsequent deliveries, and thus not all of the production total served with the British Army.

There is a story in circulation that a further batch of 60 Hercules-engined 981's was purchased from Diamond T in 1954, and that these were refurbished from ex-US Army vehicles. It is said that these vehicles were numbered in the War Office series 00 BN 00, where the normal post-war registrations for Diamond T tractors were in the series 00 YZ 00; certainly BN is a 1954/55 registration mark, and this would appear to lend the story some credence.

CHAPTER 3.1: DIAMOND T 980/981

NOMENCLATURE and VARIATIONS

Tractor, 40/50 ton, GS, Diamond T 980/981 (M20)
Hercules-engined ballast tractor, originally rated at 40 tons, and designed for use with the WW2 American-made 45 ton Rogers trailer, as well as similar units produced by other American companies such as Fruehauf and Winter-Weiss; British versions of these trailers were produced, in two 'marks' by Crane, Dysons and others. Designated M20 by the American Quartermaster Corps.

Subsequently uprated to 50 tons for use with the post-war FV3601 50 ton trailer. Available in both open and closed cab variants.

The two model designations (980 and 981) did not refer to the open and closed cab designs, as is sometimes suggested, but referred to the winch configuration; model 980 was originally described as a 'transporter' and the winch was arranged only for rear pulls, with a 91.5m long cable; the model 981 was designated as a 'recovery' tractor and the winch was installed so as to allow a front or rear pull, and the cable length was increased to 152m.

Tractor, 30 ton, GS, Diamond T 980/981 (M20)
Conversion of model 980/981 tractor to handle the 30 ton Scammell and Shelvoke & Drury semi-trailers normally fitted to the 'Pioneer'. The ballast body was removed, and a fifth-wheel turntable fitted in its place.

Tractor, 50 ton, GS, Diamond T 980-C6N/981-C6N
Post-war conversion of the 40/50 ton ballast tractor: the original Hercules engine was removed and a Rolls-Royce C6N diesel fitted in its place. Intended for use with the FV3601 trailer.

Both open and closed-cab versions were converted, but there is no evidence to suggest that any of the 30 ton fifth-wheel tractors were involved.

TRAILERS

Diamond T tractors were originally intended for use with the American-designed trailers, such as the 45 ton Rogers, but were also modified to tow the Shelvoke & Drury semi-trailer normally used with the Scammell 'Pioneer', and after the war were used with a range of standard British post-war trailers.

Ballast tractors
M9. Trailer, 45 ton, transporter, 24TW; manufactured by Rogers (model D45, LF1), Winter-Weiss (models SM 2114, 2060, 2148, 2309, and 2434), and others.
M9. Trailer, 45 ton, recovery, 24TW; manufactured by Winter-Weiss (models SM 2061, 2113).
Trailer, 40 ton, transporter, 24TW, Mk 1; known as the 'Crane design' but manufactured by R A Dyson & Co Ltd, Cranes (Dereham) Ltd, and others.

Leyland twin-blower supercharged conversion of Hercules engine (BCVM)

Air intake arrangements for supercharged Hercules engine (BCVM)

CHAPTER 3.1: DIAMOND T 980/981

Trailer, 40 ton, transporter, 24TW, Mk 2; known as the 'Dyson design' but manufactured by R A Dyson & Co Ltd, Cranes (Dereham) Ltd, and others.

FV3601. Trailer, 50 ton, tank transporter, No 1, Mk 3; manufactured by R A Dyson & Co Ltd (chassis), with Crane Fruehauf Trailers Ltd (body), (previously known as Cranes (Dereham) Ltd).

Fifth-wheel tractors
Semi-trailer, 30 ton, recovery, 8TW; manufactured by Shelvoke & Drury Ltd.

Photographs also exist showing the Diamond T coupled to the 30 ton Scammell semi-trailer.

DESCRIPTION

Engine
Hercules DXFE
The Diamond T was originally fitted with a Hercules DXFE six-cylinder, four-stroke diesel engine, with a capacity of 14.5 litres, producing 201bhp gross, and with enough torque to tear the side off a house.

The cylinder block was a mono casting, integral with the crankcase, and dry liners were fitted. The valve configuration was overhead, with both inlet and exhaust valves installed in the cylinder head, but with the combustion chamber formed to the side of the cylinder bore so as to increase turbulence of the air at the moment of fuel injection. Induction was by normal aspiration, and fuel was fed to pintle type injector nozzles by an American Bosch injector pump with an integral governor, holding the engine speed down to just 1600rpm (later increased to 1800rpm). With a relatively-low compression ratio (for a diesel), the engine was intended to run on 45 cetane diesel fuel.

A manifold heater was provided for cold weather starting, comprising a pump and spray nozzle in conjunction with a coil-energised heating electrode.

Lubrication was by conventional well-type sump.

A small, belt-driven Bendix Westinghouse compressor mounted on the crankcase was used to provide compressed air for the braking system.

Rolls-Royce C6N-143
During the 'fifties, it became obvious that the original Hercules engines were becoming beyond economic repair, and a conversion programme was initiated, replacing the Hercules with the Rolls-Royce C6NFL-143 direct-injection, four-stroke diesel.

Announced in October 1952, the C6NFL was a wet-linered in-line six of 12.17 litres, with a compression ratio of 16:1, giving a gross power output of 175bhp. The ferrous cylinder block casting was of modular design, enabling the flywheel housing to be mounted at either end, and the external accessories to be on either side (the 'L' part of the code indicating that the injector pump was mounted on the left). The valve configuration was overhead inlet and exhaust, with the valves driven by a single camshaft by means of conventional pushrods; there were two separate, removable cylinder heads.

Fuel-injection equipment was supplied by CAV, and a CAV mechanical governor held the engine speed down to 2100rpm. A separate Ki-gass (ether) carburettor was provided for extreme cold-weather starting.

Unlike the similar engines installed in the Scammell 'Constructor', in this application, the engine was a wet sump unit, with an oil capacity of 30 litres. Unusually, the engine featured both a hot-water oil heater system, and an air oil cooler. The oil heating system used engine coolant diverted from the radiator.

A small air-cooled Clayton Dewandre compressor mounted on the engine supplied compressed air for the braking system.

Engine data

	DXFE	C6NFL-143
Capacity (cc)	14,500	12,170
Bore and stroke (in)	5.625 x 6.0	5.125 x 6.0
Compression ratio	14.8:1	16:1
Firing order	-	142635
Power output (bhp):		
gross	201	175
net	78	-
Engine speed (rpm)	1600	2100
Maximum torque (lbs/ft)	685	490

Cooling system
In both applications, the engine was water cooled, using an atmospheric cooling system; the cooling water was circulated by means of a centrifugal belt-driven pump. A 610mm diameter, eight-bladed cowled fan was used to assist air circulation through the radiator and across the engine.

Transmission
On the original Hercules-engined machines, the drive was transmitted from the engine through a 368mm diameter twin-plate dry clutch manufactured by W C Lipe. On Rolls-Royce engined chassis, the clutch was a 457mm diameter single dry plate unit supplied by Borg and Beck, but retaining the American thrust bearing and release system.

The main gearbox was a four-speed and reverse unit, manufactured by Fuller, and mounted directly to the clutch and flywheel housing. A short drive shaft connected the main gearbox to the three-speed auxiliary unit, also supplied by Fuller.

The main gearbox employed constant-mesh gears for second, third and fourth speeds, with spur gears for first and reverse. The auxiliary gearbox provided three speeds, 0.77:1, 1:1 and 1.99:1, as well as incorporating a power take-off for the winch drive which allowed winching in all main gearbox ratios.

Final drive to the rear wheels was by conventional open propeller shafts, one running from the auxiliary gearbox to the first of the rear axles, with a second, shorter shaft connecting this axle to the second of the rear axles.

Axles and suspension
The front suspension consisted of a pair of multi-leaf semi-elliptical springs, clamped to the axle by 'U' bolts, and located at each end by spring brackets, shackles and pins. The axle itself, manufactured by Timken, was a conventional 'I' beam section of the reverse Elliot type, with massive wheel hubs carried on steering swivels. There were no shock absorbers but rebound rubbers were provided to control axle movement at large deflections.

The rear bogie axles were of the double-reduction type, and were similar in construction, one to another. Suspension was by means of semi-elliptical springs pivoted at their centres on a cross tube; the spring ends were supported by slip pads, allowing free movement in reaction to rebound and deflection. The axles were located by six torque rods, two above the axles, and four below. The combination of enormous differential casings and twin rear wheels effectively filled the space between the chassis rails, leaving little room for conventional axle tubes, and the Diamond T was an awesome sight from the rear.

Semi-trailer
The Shelvoke & Drury semi-trailer was fitted with an unsprung walking-beam rear-bogie, similar to that employed on the British WW2 40 ton Mk 1 and Mk 2 draw-bar trailers (see Chapter 7.1).

Steering gear
Steering was fully manual, by means of a Ross model T74 cam and twin lever system, and operated by a 560mm diameter steering wheel to give a turning radius close to 25m. The steering box drop arm was connected to the left-hand hub by a short drag link, with a full-width tie rod connecting the two hubs together.

With the enormous weight of the vehicle, and the lack of power assistance available, if you weren't a man when you got into the driving seat of a Diamond T, you probably were by the time you got out again.

Braking system
Servo-assisted compressed-air brakes were provided on all three axles; a twin-cylinder compressor installed on the side of the engine, belt-driven from the crankshaft, was used to provide the compressed air, with an indicator gauge on the dashboard.

Air line connections were installed at the front and rear of the vehicle to allow tandem operation with single-cab control, and to permit the trailer brakes to be connected to the tractor and operated via the vehicle braking system. A hand-operated reaction valve was also provided in the cab to permit independent operation of the trailer brakes, for example to prevent loss of control during steep descents.

A separate hand brake operated on a disc fitted to the output shaft of the auxiliary gearbox, and was intended only as a means of holding the vehicle at rest. The semi-trailer was fitted with a wheel-operated mechanical hand brake.

The cast-iron brake drums were 438mm diameter x 102mm width at the front, and 114mm at the rear.

Road wheels
The vehicle was fitted with single front wheels, and with duals at the rear. The wheels were manufactured by the American Budd company, and were of the detachable steel-disc type; size was 9/10.00x20in, fitted with 12.00x20 bar-grip pattern non-directional cross-country tyres. Semi-trailer wheels were 13.50x20in with standard road tyres; there were two pairs of wheels on each of the two axles.

Non-skid chains could be fitted to the front and rear wheels for additional traction. It was also possible to fit overall track-type chains around the rear wheels.

A single spare wheel was carried, in the ballast box, and on the semi-trailer, slung below the frame. The brake system compressor could also be used for tyre inflation.

Chassis
The massive heat-treated manganese-steel ladder chassis consisted of deep channel-shaped side members, running straight and parallel from front to rear, rigidly secured together and reinforced by channel-section cross members. A full-width bumper was fitted at the front, with a tow hitch installed beneath, and there was a towing pintle attached to the rear cross-member for the trailer hitch. On British Army examples, the front tow hitch was later moved to the front bumper, which was specially reinforced for the purpose.

The turntable, or fifth-wheel equipment, was of the non-detachable type, and was similar to that fitted to the Scammell 'Pioneer' tractors, and was bolted across the frame, slightly forward of the centreline of the rear driving wheels.

Photographs exist of US Army Diamond T tractors with more conventional, detachable fifth-wheel couplings, though this may well have been a post-war commercial modification.

Semi-trailer
The semi-trailer was a skeleton chassis design, with heavy steel trackways designed to support the load. The wheel configuration was four close-coupled pairs, on tandem axles.

The tracks sloped upwards towards the fifth wheel, while the rear section of the trailer was angled more sharply towards the ground to facilitate loading; the rear loading ramps were of the spring-balanced triangulated type, designed to be operated by two men. Fixed inner guide rails were attached to the trackways, and there were no outer rails, making it possible to carry oversize loads.

Winch rope guides and pulleys were fitted to permit loading with the tractor and trailer out of alignment.

Cab and bodywork
Cab
An all-steel cab was fitted, either with a non-detachable steel roof and top-hinged windscreen, or with a removable canvas roof on a tubular-steel supporting frame, together with a folding windscreen, rather in the style of the WW2 jeep. The cab was designed to accommodate a crew of two, on separate adjustable seats.

The all-steel cab was fitted with a distinctive 'peaked' roof ventilator, which incidentally was not fitted to the post-war batch of refurbished vehicles. Some of the canvas-cabbed models were provided with an anti-aircraft machine gun mount and hip ring installed on a steel framework above the passenger's seat. The normal production ratio was one vehicle off the production line with an anti-aircraft hip ring, to every nine without.

Both cabs shared the same front-end sheet metal, with an enormous 'coffin' shaped engine compartment cover and rounded front wings.

For those vehicles which were subsequently fitted with the Rolls-Royce engine, a number of modifications were made to the inner wings, and engine compartment panels. The most noticeable of these changes was the addition of a curved cover panel let into the right-hand engine compartment side cover to clear the oil filters; some vehicles were also fitted with a similar panel on the left-hand side to provide clearance for a supercharger should this be fitted at a later date. The dashboard was also modified to include a tachometer.

There were large stowage bins under the cab doors, designed to accommodate the batteries.

Rear bodywork
The rear ballast body was constructed of sheet steel, with fixed sides, and a drop-down hinged rear tailgate, designed to accommodate cast-iron ballast. Large stowage compartments were provided in the two rear corners of the body, and the front section was partitioned to provide a space for the spare wheel and for miscellaneous stowage.

A canvas enclosure, supported on a tubular-steel framework, was often fitted over the rear body to provide temporary accommodation for the crew during overnight stops, giving the vehicle a very distinctive appearance.

Electrical system
The vehicles were wired on a hybrid 24/12/6V negative-earth system, using four 6V 230Ah batteries arranged, by means of a series-parallel switch, to provide 24V for starting, 12V for charging, and 6V for lighting. The batteries were housed in lockable compartments beneath the cab.

American Delco-Remy starters and generators were fitted to the Hercules engines; but the Rolls-Royce engines were equipped with a CAV U624/21M starter and type H5512/31 CRM generator, producing 28A at 1200rpm.

The vehicles were originally equipped with American-made lighting equipment, but on many examples this was modified by the fitment of standardised FVRDE-designed British headlamps etc. The lamps were also moved from their original situation on the front wings, to a slightly-lower position on the front bumper valance.

Winch
A 40 ton capacity American-made Gar Wood winch, model 5M 723B, was mounted across the chassis behind the cab, chain-driven by a power take-off from the auxiliary gearbox.

The winch, which accommodated either 91m (model 980) or 132m (model 981) of 22.2mm steel wire rope, was controlled either from inside the cab, or from controls provided on the winch itself. Winding speed at 1000rpm with the main gearbox in first gear was 16.75m/min; an engine cut-out device was fitted in conjunction with a torque control on the winch drive shaft to prevent overloading. All of the winch controls were accessible from inside the cab and there was both an automatic brake operating on a drum on the worm gear casing, and a separate handbrake.

On the model 980, guide rollers and pulleys were provided to allow the winch rope to provide rear pulls only, while on the model 981, there were also fairlead rollers at the front.

DOCUMENTATION

Technical publications

Most of the technical documentation for the Diamond T range was produced either by the manufacturer following his own standards, or by the US Army following the post-war TM9/TM10 system. Those British documents which exist tend to deal with the re-engined models.

User handbook
Tractor, 50 ton, GS, 6x4, Diamond T 980/C6N and 981/C6N.

Maintenance manuals
Tractor, 6x4, transporter, Diamond T. TM10-1225.
Tractor, 50 ton, GS, 6x4, Diamond T. WO code 18721.

Servicing schedule
Tractor, 50 ton, GS, 6x4, Diamond T. WO code 13399.

Parts lists
Tractor, 6x4, transporter, Diamond T. TM10-1254, TM10-1224.
Tractor, 50 ton, GS, 6x4, Diamond T. WO codes 16157, 18721.
Engine, Rolls-Royce C6NFL Series 143. WO code 19900.

Technical handbooks
Technical description:
Engines, diesel, Rolls-Royce C range. EMER S532/1.
Engines, diesel, Rolls-Royce C6N: installation of type C6N 143 engine in the tractor, 50 ton, GS, 6x4, Diamond T 980/C6N and 981/C6N. EMER S532/1, supplement 1.

Repair manuals
Unit repairs: engines, diesel, Rolls-Royce C range. EMER S533.

Field and base repairs: engines, diesel, Rolls-Royce C range. EMER S534.

Modification instructions
Engines, diesel, Rolls-Royce C range. EMER S537, instructions 1-5.

Standards
Inspection standard: tractor, GS, 40/50 ton, trailer, 6x4, Diamond T. C1/86215.178, Part 1.

Base inspection standard: engines, diesel, Rolls-Royce C range. EMER S538, Part 2.

Miscellaneous instructions
Engines, diesel, Rolls-Royce C range. EMER S539, instructions 1-9.

Complete equipment schedule
Tractor, 50 ton, GS, 6x4, Diamond T. WO code 33842.

Tools and equipment
Table of tools and equipment for 'B' vehicles: tractor, 50 ton, GS, 6x4, Diamond T. WO code 17490, table 1090.

Bibliography

Army transport: data book of wheeled vehicles, 1939-1945. Fletcher, David (ed). London, HMSO, 1983. ISBN 0-11-290408-4.

Canadian Diamonds: profile of the Diamond T model 975(A) built exclusively for Canada. London, Wheels & Tracks no 24, 1988. ISSN 0263-7081.

Standard catalog of US military vehicles: 1940-1965. Berndt, Thomas. Iola, Wisconsin, Krause Publications, 1993. ISBN 0-87341-223-0.

US military vehicles, WWII. Hoffschmidt, E J and W H Tantum (eds). Boulder, Colorado; Sycamore Island Books, 1979. ISBN 0-87364-152-3.

US military wheeled vehicles. Crismon, Fred W. Sarasota, Florida; Crestline Publishing, 1983. ISBN 0-912612-21-5.

LEYLAND
CHAPTER 4

CHAPTER 4: LEYLAND

In 1896, the Leyland company opened for business as the Lancashire Steam Motor Company, constructing 30 cwt and 3 ton steam wagons.

It soon became clear that steam was not the way of the future, and the company built its first petrol-engined vehicle in 1904.

In 1907 the name of the company was changed to Leyland Motors Limited, and despite a flirtation with motorcars, Leyland tended to concentrate on bus and truck chassis, continuing with both steam and petrol engines until the former was finally dropped from the catalogue in 1926. Involvement with the military began during WW1 when 6000 of the 3 ton 'RAF type' were produced for the War Office.

The 'RAF type' continued in production until 1926, but in 1929 Leyland introduced a whole new range of chassis rated at 3.5 tons and above, and some two years later, in 1931, diversified into smaller chassis with the production of the first lightweight vehicle at the Leyland works at Kingston in Surrey. The first diesel engines were introduced in 1933.

At the outbreak of WW2, Leyland again became involved in war production, producing the 'Covenanter', 'Cromwell', 'Comet' and 'Churchill' tanks, and manufacturing engines for the 'Matilda' tank. Truck production continued with more than 1000 examples of the 10 ton 6x6 'Hippo'.

When hostilities ceased in 1945, the military orders slowed, but as part of the Ministry of Supply's standardisation plan for post-war military vehicles, Leyland constructed one prototype for the massive FV1000 60 ton tractor, and two prototypes of the FV1200 30 ton tractor.

The need for heavier vehicles had become apparent towards the end of WW2 with the increasing trend towards larger tanks. The proposed development of the FV200 'universal' tank, which ultimately led to the introduction in 1952 of the 65 ton FV214 'Conqueror' posed a serious transportation problem. Existing transporters did not offer sufficient performance or mobility, and the Ministry of Supply had developed a specification for a family of so-called 'super-heavy' 6x6 tractors. The first of these was to be a tank transporter and recovery vehicle, designated FV1003.

When this scheme was abandoned, the experience gained was put to use in the FV1200 series 30 ton tractor, planned in a variety of roles including artillery tractor, tank transporter, and recovery tractor. Although some work had been carried out by Dennis Brothers, Leyland were given the contract for the prototypes. Unfortunately nothing was to come of this project either which, had it been pursued would have put Leyland up there with Scammell as a producer of ultra-heavy chassis.

With the FV1200 project also abandoned, it was not until the introduction of the FV1100 10 ton 'Martian' chassis, also planned in artillery tractor, tank transporter and recovery tractor versions, that Leyland finally received a worthwhile production contract. Some 1300 examples of the FV1100 were ultimately produced, many remaining in service until the 1980's.

At the same time as Leyland were producing these heavy military chassis, they began the first of a series of takeovers, with the acquisition in 1951, of Albion Motors. Scammell and AEC were absorbed into the growing empire in 1955, also bringing with them Maudslay, Crossley and Thornycroft which had previously been acquired by AEC. The 1968 merger with British Motor Holdings brought the Guy and Daimler companies under Leyland's control but it must soon have become obvious that the organisation had over-reached itself. Poor management combined with a confrontational style of industrial relations led to ever more frequent crises which the Labour government of the 1970's seemed anxious to avert by frequent injections of taxpayers' money. Although the company was never nationalised, for a period it was firmly under government control.

During this time, the old marque names had gradually disappeared and production of commercial vehicles continued more-or-less exclusively under the Leyland name.

In the 1970's a number of basically-commercial 10 ton 6x4 'Bisons' were purchased by the Ministry of Defence as cargo vehicles, and the military connection continued with deliveries of the 35 tonne 'Crusader' chassis, which were being supplied under both the Leyland and Scammell names.

However, the massive Leyland empire spun, out-of-control, ever more quickly towards disaster. With the change of political climate in 1979, the cash handouts ceased, and although merger talks began with General Motors, it was the Dutch DAF truck firm which eventually took the firm over, in 1987. During this period, Leyland scored a political victory over their old rivals Bedford when the 4 tonne Leyland-DAF chassis was selected as a replacement for the trusty Bedford RL, TK and MK vehicles which had seen continuous service since the mid-1950's.

In 1993 DAF itself went into receivership and Leyland once again found itself under UK control, albeit still under the name Leyland-DAF.

CHAPTER 4.1

LEYLAND FV1000 SERIES
FV1003

The need for a range of ultra-heavy tractors had become apparent towards the end of WW2 with the development of heavier tanks. This was particularly so with the planned introduction of the 'Centurion', and the FV200 Series, which was eventually to lead to the massive 65 ton 'Conqueror'.

Unfortunately, commercial ultra-heavy truck technology was not particularly well advanced at this time, and although the trusty Scammell 'Pioneer' (see Chapter 5.1) had been successfully uprated from 20 to 30 tons during the war years, it was obvious that a further uprating to 40 tons could not be achieved without a complete redesign. Anyway, no suitable engine was available for what, at the time, was a massive payload.

A number of experiments had already been carried out in this direction. For example, before the end of WW2, some ultra-heavy draw-bar haulage trials had been carried out by FVRDE using a modified American Pacific 'Dragon Wagon' M25 armoured tractor. This vehicle, known as 'Pacific TR1 modified', was fitted with a ballast body and enormous 21.00x24 tyres. Another possible approach was to double-head two tractors to increase the draw-bar pull, and Diamond T tractors were quite routinely employed in tandem to pull/push extreme loads at that time, both by commercial operators and by the military authorities. Although this was not a particularly good everyday solution, it did enable the job to get done, and a number of excellent photographs exist, for example, of the massive 'Tortoise' tank being hauled by a pair of Diamond T's.

The lack of suitable commercial vehicles, and the outcome of the experiments with the 'Dragon Wagon' appear to have forced the War Office into believing that the only solution was to commission special designs. However, this fitted in well with the then-current thinking which was to produce a range of specialised, high-performance vehicles which would have had no commercial equivalents - the so-called 'CT' range.

Preliminary specifications for a purpose-built all-British ultra-heavy tractor, designated FV1000, were issued towards the end of the 'forties, during a period of intense activity for FVRDE. Although the FV1000 Series was initially dubbed 'super-heavy', in 1950 some wag at the War Office who clearly had nothing better to do, suggested in a memo that the series should be renamed, going on to explain that 'if a heavier tractor should ever be required, a superlative to 'super heavy' would be

TUGS OF WAR   57

CHAPTER 4.1: LEYLAND FV1000 SERIES

VEHICLE OUTLINES

FV1003

FV3301 semi-trailer

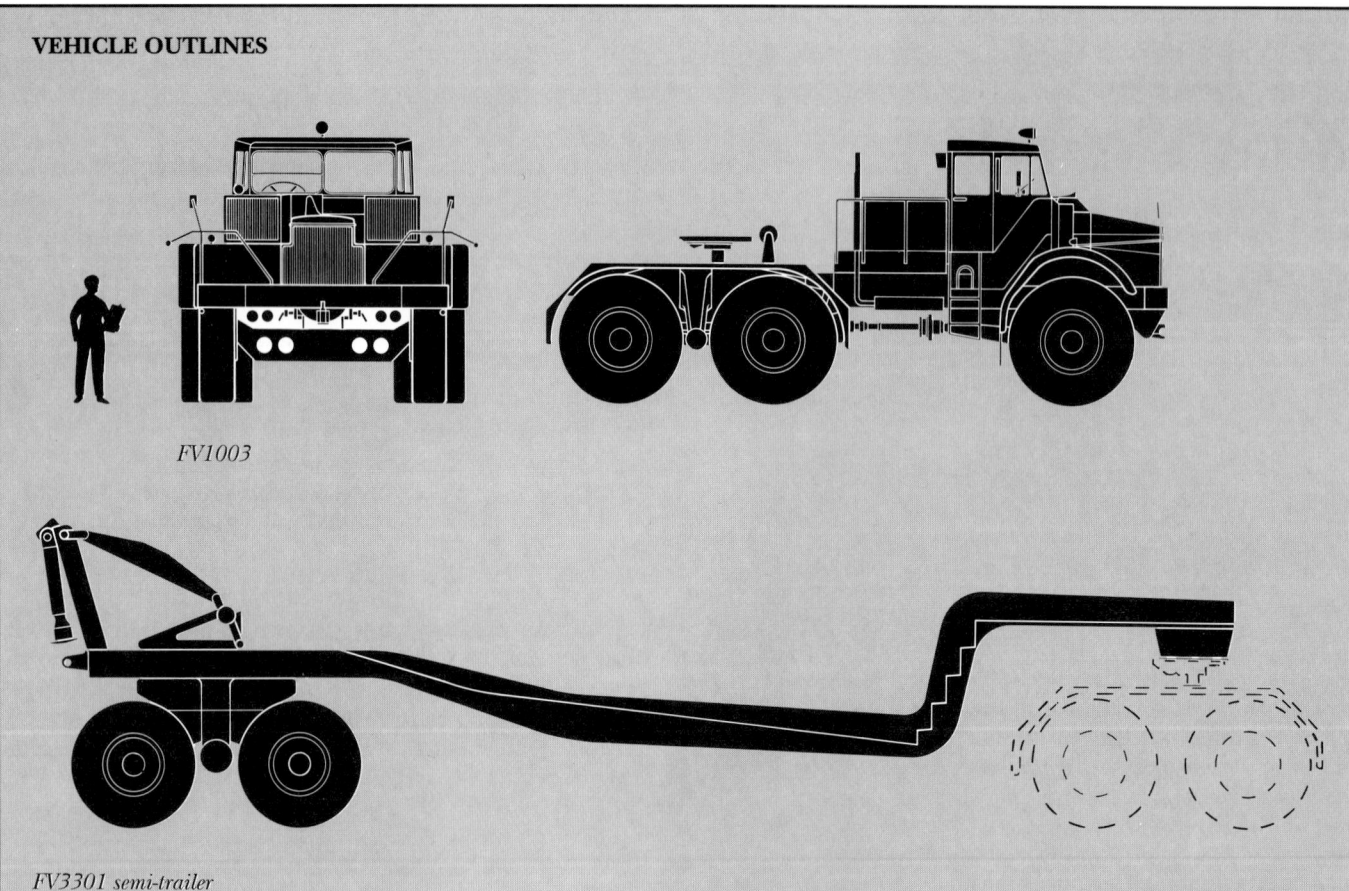

SCALE 1:100

SPECIFICATION

Dimensions
Length: chassis and cab only, 8.69m; complete tractor and FV3301 semi-trailer, 22.45m.
Width overall: 4.09m.
Height: to top of cab, 3.43m.
Wheelbase: 5.19m; rear bogies, 2.1m.
Track: front, 2.84m; rear, 2.95m
Ground clearance: 405mm.

Weight
Laden: chassis and cab only, 32,327kg; tractor and semi-trailer (with 'Conqueror' tank), 120,909kg.
Unladen: chassis and cab only, 32,327kg; tractor and semi-trailer, 56,763kg.
Bridge classification: not classified.

Performance
Maximum speed, on road: laden, 33km/h; unladen, 56km/h.
Average speed, off road: laden 25km/h.

Fuel consumption: 2.81 litre/km, on road.
Maximum range: 160km, on road.

Turning circle: tractor only, 21.3m; tractor and semi-trailer, 24.4m.
Maximum gradient: 25%.
Approach angle: 50°.
Departure angle: tractor, 45°; semi-trailer, 90°.
Side overturn angle: tractor, 45°.

Fording: unprepared, 1980mm; prepared, 1980mm.

Capacity
Maximum towed load: including semi-trailer, 88,000kg.

Engine data
Rover 'Meteorite' Mk 202A petrol injection, water-cooled V8; cylinder blocks arranged at 60° angle.
Capacity: 18,012cc.
Bore and stroke: 5.40 x 6.0in.
Compression ratio: 7:1.
Firing order: (offside cylinder block designated 'A', nearside block 'B') A4 B1 A2 B3 A1 B4 A3 B2.
Power output: 498bhp (gross); max engine speed 3000rpm; governed speed 2800rpm.
Maximum torque: 1035 lbs/ft (gross).

difficult'. It is hard to imagine that there could ever be any need for a larger vehicle!

However, no matter how it was described, the FV1000 Series would have been the largest of the post-war CT type vehicles. The tractor was designed to provide a draw bar pull of 60 tons, with the ability to deal with an all-up train weight well in excess of 100 tons. The range was planned to include a heavy tractor with ballast body for towing a full trailer (FV1001), a fifth-wheel tractor for use with a tank-transporter semi-trailer (FV1003), and a heavy recovery vehicle (FV1004), which itself was to be capable of recovering vehicles up to and including some 30 tons in weight (ie, including other vehicles from the FV1000 range).

It was originally intended that a diesel engine would be used, but no suitable units were available at the time, and the Rolls-Royce designed, Rover-built 'Meteorite' engine in petrol-injection form was used as a stop-gap measure.

A major disadvantage of large petrol engines, when compared to diesels with a similar size and performance, is that they tend to be somewhat thirsty. The 'Meteorite' was no exception, being scarcely able to better one mile per gallon, and with a fuel capacity of just 100 gallons, the FV1003 tractor had an effective range of just 100 miles. This fell far short of the FVRDE specification for the FV1000 range which had stated that 'the fuel capacity will give a range of not less than 250 miles', but of course, the proposed diesel engine would probably have brought about a significant improvement.

As it happened, it hardly mattered since the FV1000 was destined never to enter production.

DEVELOPMENT

FVRDE specifications were prepared for all of the FV1000 variants, but detailed data seems to only exist for the FV1003 tank recovery tractor version. It would appear that out of the whole range, only this version was pursued beyond the feasibility stage.

The design work was carried out at the Leyland factory in Lancashire, with the project probably getting underway during 1947.

The Ministry had indicated that there was some degree of urgency, and it was proposed that deliveries of the FV1000 Series would begin in 1948, with a maximum number of 20 vehicles being produced each year. By 1949, not only had no vehicles been delivered, but the figure was already being heavily criticised. The Ministry of Supply pointed out that the stated production figure was for one company (Leyland), and was the number of vehicles available without disrupting their normal production. Apparently, the figure could be increased in an emergency. As it happened, this was all academic since no vehicles were ever delivered at all, and it was a further two years even before the prototype was ready.

The original plan had been to use 21.00x24in single tyres at the rear of the tractor, but unfortunately, this gave a wheel loading on the rear bogie of 9.5 tons, against a previously-agreed load limit for Class 70 bridges of 6.25 tons. Although the committee responsible for such matters agreed that, in the light of the increasing weight of tanks, the Class 70 bridge classification would have to be modified anyway, there was also much discussion as to the likely effect of changing to 14.00x24in twin rear tyres. But since the combination of tractor, trailer and tank took the gross weight of the rig way over the (subsequent) overall 100 ton bridging limit imposed by the Royal Engineers, the whole project was probably doomed almost from the start.

In 1949, during the month of July, a full-size mock-up of both the tractor and trailer was produced at the Leyland 'BX' factory, complete with engine... and with a two-dimensional 'Conqueror' tank sitting on the trailer! It was photographed in the works by the Leyland staff photographer, and shows a rather tidier, and less massive treatment of the front end than that which actually appeared on the prototype vehicle. However, the sheer size of the thing must have made the staff wonder if they were reading the scale of the drawings wrongly.

The prototype vehicle finally appeared in 1951, and was a most impressive machine. With its massive appearance dwarfing anything to be found on British roads in the early 'fifties, everything about the vehicle must have seemed larger-than-life.

A special semi-trailer, designated FV3301, was also designed, and may have been produced for use with the tractor. Particular attention was paid to the off-road performance of the complete train.

However, for trials purposes, a short 'dummy', or drone, trailer was produced and loaded with ballast to simulate the all-up weight of the complete outfit. The choice of a semi-trailer, as opposed to a full trailer, was partly necessitated by the question of bridge loadings. Even if the weight limit on military bridges had been increased, the bending moment imposed on a bridge by a full trailer loaded with a 50 ton or 60 ton tank would have meant it would have been necessary to unload the tank and winch it across the bridge. The other advantage of a semi-trailer was that there would be a reduction in the gross weight due to there being no need either for a front bogie on the trailer, or for ballast over the rear wheels of the tractor.

CHAPTER 4.1: LEYLAND FV1000 SERIES

Both the tractor and the special eight-wheeled trailer were riding on huge 18.00x24in cross-country tyres. Following the discussions on bridge loading, twin rear tyres had been chosen, albeit at a compromise cross section of 21.00 rather than 14.00. This had entailed a complete redesign of the rear bogies, and brought about an increase in width to almost four metres - at the time the legal width limit for commercial trucks was just 2.3 metres. Each of these massive tyres had an effective diameter of nearly 1.6 metres, and the weight of a complete wheel and tyre was 440 kilos!

The cab roof was almost 3.5 metres high, close to the height of a double-decker bus, while the fifth wheel was more than 2 metres above the road.

The trailer was no less formidable, and where the trailer bed was raised to clear the rear wheels, the trackways were some 2.2 metres high. When loaded with a 'Conqueror' tank, the whole outfit must have presented a formidable appearance on the road, let alone when travelling across country.

With its Solex mechanical petrol-injection equipment, the 18 litre Rover V8 engine consumed fuel at the rate of 1 gallon for every mile travelled. The 500bhp gross output of the engine was transmitted through a five-speed gearbox and three-speed transfer case; the rear wheels were driven by spur gears contained in walking beams, while drive to the front wheels was transmitted through the king pins. Twin Morris radiators, 125mm thick, were mounted high above the engine where they attempted to disperse the heat generated by this prodigious rate of fuel consumption with a total cooling area of $0.5m^2$.

Although the Leyland design schedules for the period in question list four versions of the FV1000 alongside the basic chassis, it appears that only one prototype vehicle was constructed - and this was obsolete even before the trials were completed. The vehicle was civilian-registered as LYN 60, one of a series of numbers used by FVRDE on prototype and pilot vehicles. It's interesting to note that certain of these numbers seem to have turned up on more than one vehicle type - obviously the FVRDE team was not averse to moving the numbers from one vehicle to another.

The prototype machine was put through its paces at Chertsey towards the end of 1950, and FVRDE produced descriptive data sheets in late 1951 for both the chassis-cab, and the complete vehicle, so presumably still believed that the project might be progressed further. A photograph of the FV1003 prototype was also included in a publicity leaflet which Chertsey had put out in the early 'fifties, possibly to publicise the first British military

FVRDE-modified 'M25 Pacific TR1' with ballast body (IWM)

Full-size mock-up of FV1000 tractor at Leyland factory, 1949 (BCVM)

Wooden tank mock-up on FV3301 semi-trailer, 1949 (BCVM)

CHAPTER 4.1: LEYLAND FV1000 SERIES

Excellent view of mock-up for complete tractor and semi-trailer (BCVM)

Complete working prototype of FV1000 tractor, 1951 (IWM)

FV1000 tractor dwarfs FV1800 Mudlark vehicle (TMB)

vehicles exhibition, and several photographs exist of the FV1000 and FV1200 prototypes side by side.

However, the combination of the weight, together with mobility and cost problems proved insurmountable and ultimately nothing was to come of the project. A War Office memo produced at the time suggested that the trials were only continued to provide experience for the FV1200 series (see Chapter 4.3).

Ultimately, the problem of transporting the 'Conquerors' was solved at a far more reasonable cost by the introduction in 1951, of the FV12001 Thornycroft 'Antar' (see Chapter 6.1), which although only rated at 50 tons, was more than equal to the task originally assigned to the FV1003. In March of 1955, it was stated that the FV1000 project had been 'entirely dropped'.

So, there were no orders forthcoming and no further interest was taken in development of the prototype, but the military authorities did try to encourage Leyland to market the vehicle elsewhere. Whilst it is hard to imagine who, apart from the military, would have need of a 60 ton tractor with an overall width of nearly four metres, as far back as 1948 the War Office had categorically stated to Leyland that the FV1000 was 'not subject to any security classification ... (and it)... would be advantageous if Leyland were to exploit the vehicle with civilian firms... without stating the Army's interest'. Not surprisingly there were few takers!

The prototype vehicle was to remain at Chertsey until the mid 'seventies, where it was used for gradient simulation duties, but it's a pity that this monster was never to really turn a wheel in anger. It is possible that, like the two FV1200 prototypes, the tractor ended up in Hardwick's yard at Ewell but no trace of it remains today.

NOMENCLATURE

FV1003. Tractor, 60 ton, CT, heavy, 6x6, Leyland.

There were no variations: although other variants were specified, none appear to have been produced.

TRAILERS

There was a purpose-designed trailer intended for use with the FV1003, and expressly intended for off-road use, though whether or not any were produced is hard to say:

FV3301. Semi-trailer, 60 ton, tank transporter, cross-country.

CHAPTER 4.1: LEYLAND FV1000 SERIES

DESCRIPTION

Engine

The original intention had been to fit a diesel engine but it seems that none was available at the time with sufficient power, so the decision was taken to employ the Rolls-Royce designed Rover 'Meteorite' engine.

This was an overhead-valve V8 petrol engine, effectively configured as two-thirds of the hugely successful Rolls-Royce V12 'Merlin' and 'Meteor' engines which had been employed in both tanks and aircraft during WW2. Unfortunately, the 'Meteorite' was designed to run on 80 octane fuel where the standard military vehicle fuel in use in 1950 was rated at only 70 octane. Despite the logistical supply problems which this was likely to present in combat situations, the additional power available was felt to be sufficiently important, and the decision was taken not to make any design modifications to the 'Meteorite' to allow it to run on the lower grade fuel.

The version used in the FV1000 was fuel-injected, employing four valves per cylinder. The engine was of dry-sump configuration with an oil reservoir of 45 litre capacity; a spiral finned-tube oil cooler was fitted. Fuel-injection equipment was supplied by the SU Company. The fuel was delivered to an SUX 727 eight-point injector pump by means of a David 'Korrect' fuel pump mounted on the block and driven by the camshaft.

The ignition was of the magneto type, employing twin British Thompson Houston (BTH) Type C8B magnetos producing 12kV.

Cooling system

The engine was water cooled, with the cooling system pressurised to 7050kg/m². The cooling water was passed through a pair of Morris 'H' matrix radiators with 10 rows of tubes, giving an overall thickness of 125mm. A massive 10-bladed cooling fan of some 550mm diameter was installed behind each radiator and driven at crankshaft speed by means of a shaft and bevels through a slipping clutch.

Although the vehicle was provided with an air inlet ahead of the engine in the conventional position, there were also what appeared to be air intakes mounted either side of the cab, almost at windscreen height, with sheet metal ducts running outside the cab, apparently designed to discharge the air below and behind the crew compartment. These inlets and ducts may well have been provided for the engine oil coolers, or alternatively may have been intended for the cooling system radiators, with the oil coolers installed at the front of the vehicle.

Transmission

Driving through a 560mm hydraulic coupling and friction

Side elevation shows fifth wheel and rear bogie arrangements (IWM)

Ballast trailer mock-up to simulate weight of FV3301 (REME)

FV1000 and FV1200 tractors - combined width 7.19m! (IWM)

CHAPTER 4.1: LEYLAND FV1000 SERIES

Rear view of FV1000 clearly shows the colossal width across tyres (IWM)

FV1000 tractor on road trials - note size of Diamond T behind (IWM)

... imagine meeting this on a blind bend (IWM)

clutch, power was transmitted via a five-speed and reverse Leyland gearbox, to a separately-mounted three-speed transfer gear assembly. Selection of the transfer gears was servo-assisted. The main gearbox also provided a power take-off to drive a mechanical winch.

From the auxiliary gearbox, open propeller shafts ran the length of the vehicle to drive the front and rear axles. The transfer box ratios were 0.8093:1, 1:1 and 2.002:1.

Drive to the rear axles was by means of bevel-and-spur reduction gears to the walking-beam centres, with spur drive to the wheels through the walking beams.

Axles and suspension
Suspension was provided for the front axle only, and no shock absorbers were fitted. A single transverse elliptical spring was installed across the chassis at the front axle, pivoted centrally in a swivel housing designed to handle movement of the road wheels. The total suspension movement from bump to rebound was 200mm.

There was no rear suspension, but bump stops were fitted to limit the gear casing movement to 305mm either up or down as the walking beam pivoted to accommodate undulations of the terrain.

The solid front axle casing incorporated a bevel-gear type differential from which power was transmitted, via axle shafts and bevel gears, to the road wheels. In order to gain maximum ground clearance at the front, the drive was passed vertically through the swivel (king) pins, with the top bevel gears transmitting drive from the axle to the vertical pin, and the bottom gears transferring the power to the hub, and thence to the road wheel.

Semi-trailer
The trailer was provided with a single, unsprung walking-beam bogie, with each axle carrying eight wheels.

Steering gear
Steering was controlled through a hydraulically-assisted worm-and-sector type steering box.

A massive 560mm diameter steering wheel was fitted, giving 4.6 turns from lock to lock, with power assistance. The FVRDE data sheet, dated October 1951, also stated that without power assistance the steering wheel required 14.6 turns from one lock to the other! Whether this means that the power assistance was optional is unclear, but it is hard to imagine driving a vehicle where even the slightest deviation from straight ahead must have required several complete turns of the wheel, so perhaps this was a typing error in the data sheet.

Braking system
All three axles were provided with twin-leading shoe, dual line air-operated brakes, actuated by a foot pedal

with servo assistance from piston-operated air cylinders. In addition, there was an engine exhaust brake.

The cast-steel brake drums were 560mm diameter, with shoes of 225mm width on the rear wheels, and 112mm width on the front.

The hand (parking) brake operated via a cross shaft acting mechanically on a drum fitted to the rear of the auxiliary gearbox.

No definitive data seems to exist but presumably the trailer brakes were connected to the tractor air lines.

Wheels and tyres
Wheels were 13.00x24in, with demountable rims retained by bevel clamps on a cast felloe. The tyres were Goodyear 'All Service' NDCC type, 18.00x24 (the FVRDE data sheet for the vehicle produced in 1951 stated that the wheels were 29in diameter but it is likely that this was also simply an error).

Non-skid 'overall track' type chains could be fitted to all wheels: each chain weighing in the region of 750kg.

Although the FVRDE specification called for 'a spare wheel and tyre... (to) be provided in an easily accessible position', it does not appear that any such provision was made on the prototype vehicle.

Chassis
The chassis was of channel and box section design, running more-or-less straight from front to rear; a full width, channel-section bumper was fitted at the front.

The fifth-wheel platform was mounted slightly forward of the tractor bogie centreline, and both pin and pintle type towing attachments were provided, rigidly mounted at the front and rear.

Semi-trailer
The proposed semi-trailer was a massive affair consisting of a pair of steel runways on a girder chassis riding 915mm from the road at the lowest point. It hardly matters whether this trailer had a standard fifth-wheel coupling, since it was sited 2135mm above the road, and there is no way it could have been used with any other tractor.

In order to maintain a sensible centre of gravity, the tank was carried in a nose-down attitude towards the fifth wheel of the tractor. Loading was effected by means of three-part triangulated ramps at the rear.

The FVRDE documentation suggests that there was seating for one person on the semi-trailer; unfortunately, the drawings do not show where this might have been.

Cab and bodywork
Nobody could argue that the body of the FV1000 was a thing of beauty, but what it lacked in aesthetics, it certainly more than made up in sheer presence.

Although the massive front wings were gracefully curved, most of the body was flat, consisting largely of simple panels with strange-shaped infill pieces wherever the panels from different planes didn't quite come together. The actual crew compartment was of more-or-less normal width, designed to house a crew of three, and looking somewhat incongruous perched on the colossal width of the chassis. The windscreen was the standard two-piece design of the period, with the driver's side divided into fixed and opening lights.

The doors were positioned some 1500-1700mm above the road, with access provided by tubular metal ladders. Tubular rails were provided around the front wings and bumper, presumably to form a guard to what were effectively walkways, as well as providing a mounting for a number of strategically-placed tiny rear-view mirrors.

The winch was housed in a separate compartment between the cab and the fifth wheel coupling and protected at the rear by tubular guard railings.

The rear wheels were covered by simple flat panelled splash guards with side valances.

Electrical equipment
The electrical system was rated at 24V, and wired negative earth; all wiring and electrical devices were screened against radio interference. Four 12V 60Ah batteries were fitted, charged by an engine-driven CAV generator rated at 40A maximum output.

Winch
The prototype vehicle was fitted with a 20 ton capacity mechanical winch with rope fairleads and pulleys installed in such a way as to allow either front or rear pulls. The specification called for 137m of winch cable to be installed, with the winch speed to be controlled by the hand throttle; the minimum speed was 4.6m/min, with the maximum not to exceed 15.25m/min.

Radio equipment
The FVRDE specification called for provision to be made for installing 'Larkspur' C40 or B40 sets, 'or subsequent equivalent'; there was no sign of a radio being fitted in the prototype vehicle.

DOCUMENTATION

FV Range of vehicles. Information brochure: Part 2, 'B' vehicles. FVRDE, October 1951.

CHAPTER 4.2

LEYLAND MARTIAN FV1100 SERIES
FV1103, FV1119, FV1122

At more-or-less the same time as becoming involved in the abortive FV1000 and FV1200 heavy tractor projects, Leyland were also asked to undertake development of the more modest FV1100 chassis, popularly known as 'Martian'. Despite some initial difficulties, this vehicle was considerably more successful than the FV1000/FV1200, going on to be produced in seven variants, with some 1300 production examples.

Rated at a nominal 10 tons, the 'Martian' range was intended to include tractors for the 10/20 ton semi-trailer (FV1101), a medium/heavy artillery tractor (FV1103), and a heavy recovery tractor (FV1119). There were also planned cargo, fuel tanker, crane, workshop, and command/signals variants, though of these, only the cargo vehicle was pursued beyond the planning stage.

The first prototypes, produced in 1951, were of the cargo variant, probably intended to replace the ageing WW2 AEC 'Matadors', and Leyland 'Hippos'. However priorities must have been changed, and the first production examples were of the medium artillery tractor. Subsequently, there was also a heavy recovery vehicle, intended to work alongside the GS-rated Scammell 'Explorer (see Chapter 5.3), and a special artillery tractor for the 8in 'atomic' howitzer.

Equipped with an eight-cylinder Rolls-Royce 'B Series' petrol engine which provided enormous reserves of low-down power, the result was a vehicle with a modest turn of speed, but with considerable pulling power. The prototypes and the first 25 vehicles produced were powered by a B80 engine, but this was subsequently replaced in production by the more-powerful B81, and the earlier vehicles were subsequently upgraded.

Excellent off-road performance was achieved by the use of a Scammell-style centrally-pivoted front axle suspended on a transverse semi-elliptical spring. The rear suspension and drive train arrangements also broadly followed the Scammell design, employing a gear-driven, centrally-pivoted walking beam suspended on longitudinal semi-elliptic springs. Of course, by the time these vehicles were constructed, Leyland had acquired Scammell so perhaps the resemblance in so many areas should not come as a surprise.

Drive to the front wheels, when required, was transmitted vertically through the king pins, a system previously employed on the WW2 Mack trucks.

CHAPTER 4.2: LEYLAND MARTIAN FV1100 SERIES

In all, there were seven military variants, and in 1954, Leyland announced a diesel-engined commercial variant using what was virtually the same chassis.

Despite considerable problems with unreliability in the early vehicles, the 'Martian' eventually proved itself to be a well-built and impressive truck intended for heavy and severe cross-country conditions.

DEVELOPMENT

The 'Martian' was developed as part of the family of CT-rated vehicles planned by FVRDE towards the end of WW2. The first variant to be prototyped was the cargo vehicle, and at least one example was produced in 1951, costing the Ministry £8550. This prototype was not a resounding success, and was described in a War Office paper of 1955 as 'being subject to so many failures... that it had to be returned to the makers for reworking'.

A commercial cab design was used, looking like it might have been based on the contemporary 'Hippo'. The choice of moving 'cycle wings' meant that the lower corners of the cab had to be cut back to provide sufficient clearance for the wheels and wings when the axle was on full articulation. This contrived to give the cab and wheel arch area a curious cut-off, unfinished look. This was changed on the production vehicle, which featured fully-enclosed pressed-steel front wings, providing sufficient internal clearance for the wheels on full travel.

Although the Ministry had contracted the work to Leyland, one commentator writing in the early 'fifties stated that the prototypes were constructed by Thornycroft. No reliable documentary data appears to exist to support this view, but by chance a photograph taken in the AEC works in the early 'fifties shows what are clearly two 'Martian' prototype chassis in the corner of the workshop. Thornycroft, of course, had been taken over by AEC in 1961, and there may well have been some co-operation between the two firms before that time. So the story could have some credibility.

Four prototype cargo vehicles were initially issued for what proved to be a very short trials period; registration numbers 47 BM 70, 72, 96 and 98.

Artillery tractor

The second variant to be produced was the FV1103 heavy/medium artillery tractor, which was intended to replace the WW2 AEC 'Matadors', and was designed for towing the 5.5in howitzer. Ultimately the 'Martian' was also developed into variants intended to pull the 7.5in and 8in howitzers, 40mm anti-aircraft guns, and other field pieces; and the vehicle was also assigned to replace the Mack tractors which were being used with the 7.2in

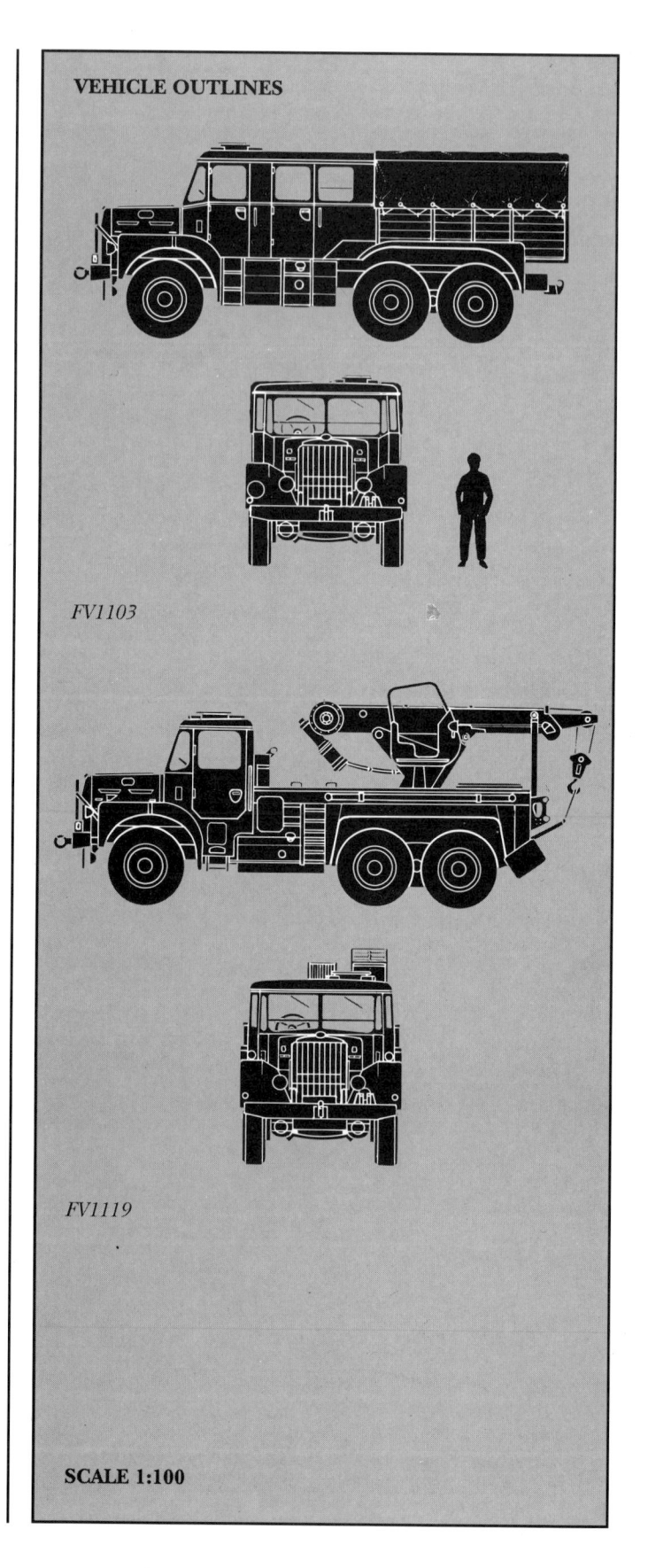

VEHICLE OUTLINES

FV1103

FV1119

SCALE 1:100

CHAPTER 4.2: LEYLAND MARTIAN FV1100 SERIES

SPECIFICATION

Dimensions and weight

	FV1103	FV1119	FV1122
Dimensions (mm):			
Length	8185	8890	8380
Width overall	2591	2591	2591
Height:			
to top of cab	3073	2990	3073
to top of tarpaulin	-	-	3580
over crane	-	3200	-
load platform	1384	-	1711
Wheelbase	4420	4420	4420
Track:			
front	2089	2100	2089
rear	2096	2100	2096
Ground clearance	465	465	465
Weight (kg):			
Laden	18,660	-	24,600
Unladen	14,310	21,630	14,310
Bridge classification	24	22	23
Crane dimensions (mm):			
Reach:			
minimum reach	-	3050	-
maximum reach	-	5490	-
Maximum lift height	-	8235	-
Maximum slew, left/right	-	120°	-
Max load (kg)	-	15,000	-
Load at max reach (kg)	-	1500	-

Performance

Maximum speed, on road: 42km/h (FV1103/FV1122); 56km/h (FV1119).
Average speed, off road: 25km/h.

Fuel consumption: 0.81 litre/km, on road.
Maximum range: 485-560km, on road.

Turning circle: 21.34m.
Maximum gradient: 33%.
Maximum gradient for stop and restart: 26%.
Approach angle: 40°.
Departure angle: 36° (FV1103/FV1122); 42° (FV1119).

Fording: unprepared, 760mm; prepared, 1980mm.

Capacity

Maximum load: 10,000kg.
Maximum towed load: 7500-8000kg.

---

howitzer, a role originally intended for the FV11002 AEC 'Militant'.

As well as being designed to tow the gun, both on and off road, the tractor was also expected to have sufficient reserves of power, 'by means of engine and winch', to manoeuvre the 5.5in howitzer into and out of gun pits etc.

A specification for the vehicle was issued as early as 1948, and was extensively revised in 1950 when an early prototype chassis, with a B80 Mk 2H engine, was rebuilt to as nearly as possible represent the chassis which were to be produced on the production lines, and passed to FVRDE for trials. The earlier prototype of the cargo vehicle had not been given a proper trial but had been subjected to what were described as 'modified user trials', which were scheduled to last just five days. It was pointed out at the time by the War Office that these trials were not intended to replace 'normal' user trials, which would have lasted six months, but had been forced onto them by 'circumstances outside of their control'. You might think they would have learned something from this experience.

Not a bit of it! In October 1953, at an acceptance meeting held at Larkhill, the Director of Weapons Development (DWD) and Deputy Director, Royal Artillery (DDRA) accepted the vehicle, subject to certain modifications, after a test covering less than 10,000km, and before user or troop trials had been conducted. Specific mention was made at the acceptance meeting of problems with the power steering system, which were to be rectified by the use of an improved type of hydraulic ram after vehicle number 151. Other changes which were to be incorporated included a revised front bumper, inclusion of hand throttle control, changes to the winching pulleys, and various amendments to the body and body mountings. On the prototype vehicles, the air cleaner was mounted outside the bonnet in the style of the Scammells.

By late 1953, production of the artillery tractor had begun, with the average price per vehicle contracted at £7850.

The first 25 vehicles were fitted with a B80 engine. This was then replaced in production by the more-powerful B81 and the intention was that the earlier vehicles would be upgraded.

However, it could hardly have come as a surprise that all was not well. As early as December 1953, Royal Artillery units were reporting defective clutches and transfer gearboxes. After a short period in service, the vehicles were withdrawn while FVRDE tried to find and rectify the source of the clutch problems. Two tractors (43 BM

42 and 43) were subjected to a four-month long reliability trial by BAOR towards the end of 1955, and a final trials report was issued in January 1956. Despite the clutch on one vehicle collapsing completely after just 2400km, only minor comments were made and the vehicles were generally given a clean bill of health.

In 1957, vehicles were re-issued and reported to be mechanically sound but by May 1958, the Director, Royal Artillery (DRA) was once again requesting that the vehicles be withdrawn from service and replaced by the 10 ton AEC 'Matador' tractors, of which there was apparently a surplus. In June of 1958, a Major of the General Staff wrote that 'this vehicle has had an unsavoury history', stating 'I do not believe we should accept this vehicle back into service until exhaustive troop trials have been carried out... after modification.'

It was not until October 1958 that the Ministry of Supply claimed to have found a final solution to the clutch problem and was suggesting that the vehicles be returned to REME workshops for modification during the 1959/60 financial year. Finally the 'Martian' was able to earn its keep and the Royal Artillery began to look into other possible roles for the vehicle.

In August 1964, FVRDE issued a report into the suitability of using a modified FV1121 cargo vehicle to replace the ageing WW2 Mack 6x6 tractors, designed to tow the American 8in 'atomic cannon' howitzer without a limber. FVRDE had carried out a number of chassis and body changes to a basic short-wheelbase cargo truck (47 BM 31), using the maximum number of towing components taken from the Mack, with the idea of testing its suitability for this role. The trials were carried out over a distance of some 800km, and the report concluded that the vehicle would 'satisfactorily tow the 8in howitzer without a limber and could be used as a replacement for the Mack'. Production drawings and specifications were prepared and the resulting tractor, which looked very much like the cargo vehicle on which it was based, was designated FV1122. There is some evidence to suggest that, like the FV1121 cargo vehicle, this might have been a conversion of obsolete FV1103 vehicles.

Recovery vehicle

Development of the recovery vehicle arose out of the serious shortage of heavy recovery vehicles which existed in the immediate post-war years.

This shortage remained until well into the 1960's in spite of the delivery of the Scammell GS range 'Explorer' FV11301 recovery vehicles in 1950. The recovery version of the 'Martian' family was an attempt to address this shortage, whilst also providing a CT-type vehicle with enhanced cross-country capability.

Prototype FV1100 demonstrates front axle articulation (BCVM)

Early prototype on test at FVRDE - note canvas side screens (IWM)

Prototype cargo vehicle with full-height doors (IWM)

CHAPTER 4.2: LEYLAND MARTIAN FV1100 SERIES

Prototype chassis on Leyland's articulation gauge (BCVM)

Early production example of FV1103 - note poor panelwork (BCVM)

Definitive example of FV1103 on test at FVRDE (IWM)

Work on the FV1119 recovery version began in the mid 1950's, with production vehicles being issued to units, at a cost of £22,500 each, during 1962/63. Presumably by the time the recovery vehicles were issued, the problems which had been experienced with the artillery tractors had been overcome.

In 1960, the Department of the Army in Australia had agreed to undertake tropical trials on two UK vehicles. One of these was a prototype or early production example of the B81-equipped FV1119 recovery tractor, the other was the FV601 'Saladin'. At the same time, similar trials were conducted on an Australian-built Land Rover ambulance, and a Canadian tracked Bombardier 'Muskeg'. The trials, which began on 2 November 1960, covered 16,000km, and included a hot-wet phase in Northern Queensland, and a hot-dry/dusty phase in Western Queensland, Northern Territory, and South Australia. The object of the trials was to assess the suitability, performance and reliability of the vehicles under tropical conditions, to assess the deterioration of the vehicles and components when parked in the open, to obtain data for future vehicle design, and to ascertain any special servicing or maintenance requirements for vehicles used in these conditions.

Despite a number of failures, including the steering hydraulics, unloader valve, compressor drive shaft, and exhaust pipe supports, the trials report concluded that both the performance and reliability of the vehicles was satisfactory. The recovery tractor was considered suitable for operation under severe tropical conditions although some adverse comment was made regarding crew comfort.

The total cost of the trials was £10,000, and the vehicles were shipped back to the UK in July 1961.

Production

All of the production chassis were constructed by Leyland Motors at their 'Ministry of Supply' factory in Lancashire, with production taking place over a roughly 15 year period spanning the early 'fifties through to the mid 'sixties. No bodies were produced by Leyland, and the completed, running cab-and-chassis assemblies were delivered to the bodybuilder to be equipped with the appropriate body.

There is evidence to suggest that at least some of the recovery vehicles were converted from cargo vehicle chassis, and it is possible that some un-bodied, but otherwise completed chassis were put into store.

The bodies for the FV1103 artillery tractor were coach-built either by Park Royal Vehicles at their North London works, or by Mann-Egerton at Norwich.

CHAPTER 4.2: LEYLAND MARTIAN FV1100 SERIES

The recovery equipment for the FV1119 was designed and built by Royal Ordnance Factories working with Leyland Motors.

The total number of 'Martian' vehicles produced was 1380; this includes slightly less than 60 of the FV1103 artillery tractor, and 280 recovery vehicles. Although this leaves around 500 cargo vehicles, it must be remembered that a number of FV1103 tractors were apparently converted to cargo and howitzer tractor configurations.

In 1954, Leyland offered the 6x6 'Martian' chassis to the commercial truck market. Equipped with the popular Leyland 0600 diesel engine, the vehicle was announced in 8 ton tractor form, with a wheelbase of 7575mm, at the 1954 Commercial Motor Show; an 8481mm wheelbase 10 ton load carrier was described as 'soon to be available'.

NOMENCLATURE and VARIATIONS

FV1103(A). Tractor, 10 ton, CT, medium artillery, 6x6, Leyland, Mk 1
Towing and supply vehicle for medium artillery (up to 8 ton gross weight). Fitted with insulated crew cab for 12; a small timber-sided cargo body allowed up to 5000kg of ammunition to be carried behind the cab. Fitted with a 10 ton mechanical winch, and with a small chain hoist in the rear body to simplify handling of the spare wheel, and to enable the gun trail to be lifted to the tow hook.

FV1119(A). Tractor, 10 ton, CT, recovery, heavy, 6x6, Leyland, Mk 1
Heavy recovery tractor fitted with hydraulic crane having an extending and slewing jib, and with a 15 ton two-speed hydraulic winch installed under the cab, configured for front and rear pulls. Intended for recovery and suspended tow of wheeled and tracked vehicles up to 10 ton weight; also designed for use with the FV3221 10 ton light recovery trailer.

FV1122. Tractor, wheeled, 10 ton, GS (8in howitzer), 6x6, Leyland, Mk 1
Modified version of FV1121 cargo truck, intended as a towing and supply vehicle for the 8in howitzer, either with or without the appropriate limber; body and cab designed to accommodate an 11-man crew. A chain hoist was incorporated to enable the gun trail to be lifted to the towing position.

TRAILERS

The FV1119 recovery tractor was intended to be used with the 10 ton light recovery trailer, which had been designed for recovering and transporting CT and GS category wheeled vehicles up to 10 tons in weight, and the FV430 Series tracked vehicles:

Chain hoist used to lift gun trail and to handle spare wheel (IWM)

FV1103 on trial at FVRDE with 5.5in howitzer (IWM)

Six of the gun crew manhandle the howitzer into position (IWM)

CHAPTER 4.2: LEYLAND MARTIAN FV1100 SERIES

Early production example of FV1119 shows its pulling power (BCVM)

Crane production line - final assembly carried out at Leyland (BCVM)

Early production FV1119 with crane operator's cover in position (BCVM)

FV3221. Trailer, 10 ton, 4TW/2LB, recovery, light; manufactured by Crossley Motors Ltd, Rubery Owen & Co Ltd, and J Brockhouse & Co Ltd.

Photographs also exist showing the artillery tractor with various types of, usually low-loader trailer, but this would have been an expedient rather than a regular occurrence.

DESCRIPTION

Engine

The 'Martian' was fitted with a standardised Rolls-Royce 'B Series' eight-cylinder petrol engine. The 'B Series' consisted of a range of closely-related four-, six- and eight-cylinder engines designed to incorporate a high degree of commonality of components, and purpose-designed for military use.

The 5675cc B80 Mk 2H version was used for the first 25 artillery tractors, but all other variants were fitted with the bored-out B81, either in Mk 2H, Mk 5H (most production vehicles) or Mk 5K (some recovery tractors) form, with a capacity of 6516cc.

The B80/81 engines were normally-aspirated water-cooled, in-line eight cylinder units with overhead inlet valves and side exhaust, with both the monobloc crankcase and detachable cylinder head of cast iron. Pressure lubrication was by means of a conventional wet sump. Fuel was supplied by a David 'Korrect' mechanical diaphragm-type pump to a dual-downdraught Solex 40NNIP carburettor, and Ki-gass cold-start equipment was installed on the water-jacketed inlet manifold.

Although the engines were both thirsty and expensive, they offered a very high degree of reliability with a life in use, which often exceeded 300,000 kilometres.

Engine data

	B80 Mk 2H	B81 Mk 2H, 5H or 5K
Capacity (cc)	5675	6516
Bore and stroke (in)	3.50 x 4.50	3.75 x 4.50
Compression ratio	6.4:1	6.4:1
Power output (bhp):		
gross	165	220
net at 3750rpm	136	195
Maximum torque (lbs/ft):		
gross	280	330
net at 2250rpm	257	315
Firing order	16523874	16523874

Cooling system

The engines were water-cooled in the conventional manner by pump-assisted thermo-siphon, with the

CHAPTER 4.2: LEYLAND MARTIAN FV1100 SERIES

radiator pressurised to around 7050kg/m². The cooling fan was driven by tandem belts.

Transmission
Driving through a 280mm Borg and Beck twin dry-plate clutch, power was transmitted through a four-speed and reverse gearbox, to a three-speed transfer gear assembly housed in a unit-constructed auxiliary gearbox. First and reverse were engaged by a sliding gear, whilst the other four speeds were of the constant-mesh helical type. The auxiliary gearbox also provided drive for the winch, or winch hydraulic motor.

From the auxiliary gearbox, open propeller shafts ran the length of the vehicle to drive the front and rear axles. A third, shorter shaft was used to drive the winch or winch pump where applicable. Engagement and disengagement of the front axle was effected by a sliding coupling sleeve, with the mechanism installed within the auxiliary gearbox casing, and controlled manually by a lever in the cab. The auxiliary gearbox also included sliding dog speed change and winch control facilities, operated by means a third centrally-located lever in the cab.

The transfer box ratios were 1.055:1, 1.360:1 and 2.690:1 for the axle and (mechanical) winch drives, and 1:1 for the winch hydraulic pump power take-off on recovery vehicles. Low auxiliary gear was interlocked through the all-wheel drive system, and could not be engaged unless the front-axle drive was also selected.

In line with War Office policy of the time, it was originally intended that a central tyre-inflation system would be used, allowing the tyre pressures to be altered while the vehicle was moving. Although the valve housings were installed on the hub, the system was never completed.

Suspension and axles
Suspension was provided at three points by means of semi-elliptical springs. No shock absorbers were fitted.

A single spring with 14 leaves, installed across the chassis at the front axle, was pivoted centrally in a rubber-mounted swivel housing, designed to handle movement of the road wheels. The right-hand end of the spring was attached to the torque frame by a shackle spring, the other end resting on a pivoted rocker.

The rear suspension consisted of two longitudinal main springs, each with 10 leaves, together with additional helper springs on the recovery vehicle, installed on each side of the chassis. The axle shaft casing was attached to the springs by U bolts, with the walking-beam gear casings designed to pivot on the axle casing. A mechanical locking device was provided on the rear suspension of the recovery vehicle to assist in lifting operations. Torque-

Same vehicle from front (BCVM)

... and here the FV1119 lifts 18 tons dead weight! (BCVM)

Production example of FV1122 artillery tractor (TMB)

reaction arms were installed in 'Metalastik' bushes on the axle and chassis frame.

Bump movement of both the front and rear axles was limited by rubber stops on the chassis frame. Total suspension movement from bump to rebound was 305mm at the front, and 255mm at the rear.

Drive to the single rear axle casing was transmitted to a bevel-gear type differential and then through axle shafts and spur-type reduction gears to the four road wheels. The reduction gears were housed in walking-beam gear cases located on either side of the chassis; the cases being designed to articulate on central pivots attached to the chassis members.

The solid front axle casing incorporated a bevel-gear type differential from which power was transmitted, via axle shafts and bevel gears, to the road wheels. In order to gain maximum ground clearance at the front, the drive was passed vertically through the swivel (king) pins, with the top bevel gears transmitting drive from the axle to the vertical pin, and the bottom gears transferring the power to the hub and road wheel. The axle and bevel-gear housings were mounted in a torque frame which had its rear end attached to the main chassis members by means of ball ends located in spherical bearings.

The front axle was provided with a very-substantial tubular guard, running either straight across from one side to the other (on prototype vehicles), or curved downwards in the centre to protect the differential. On some photographs of prototype vehicles, this guard could be mistaken for an undriven axle tube.

Steering gear
Steering was controlled through a hydraulically-assisted cam-and-roller type steering box.

Steering action from the wheel was transmitted through the steering column, via bevel gears at its base, and through a short prop shaft to a relay box which incorporated a cam-and-roller mechanism. A drop arm from the relay box was connected by a short drag link to a triple-armed relay lever pivoted on the front torque frame; steering rods were used to connect the relay lever to the front wheel swivels. A third arm was used to connect the relay lever to a hydraulic ram designed to provide power assistance.

There were two designs for the hydraulic system according to the type of air/hydraulic accumulator used. In the 'Series 1' design, fitted to the first 286 vehicles, either a single air-bag or a single piston-type accumulator was fitted, while the 'Series 2' employed two piston-type accumulators.

Both systems operated in broadly the same way. The action of turning the wheels from the straight-ahead position opened a steering control valve in the hydraulic circuit; this allowed hydraulic fluid to be pumped under pressure to the hydraulic ram which imparted the appropriate movement to the steering relay lever, either pushing or pulling according to which way the steering wheel was turned. As soon as the wheel was turned in the opposite direction, the action of the ram was reversed and assistance was provided to return the wheels to the straight-ahead position.

A massive 560mm diameter steering wheel was fitted, giving 6.33 turns from lock to lock.

Braking system
All six wheels were provided with twin-leading shoe air-operated brakes, actuated by a conventional foot pedal. In addition, there was a hand-operated trailer braking system and a separate hand-operated 'hill-holder' system designed to apply all of the brakes.

Air pressure was generated by a twin-cylinder compressor, gear-driven from the engine power take-off, and maintained in two chassis-mounted reservoirs. The braking system air was drawn through a combined air cleaner and anti-freeze device, designed to prevent the braking system from icing up. A gauge on the dash indicated braking system air pressure to the driver.

The cast-iron brake drums were 394mm diameter x 118mm width on the rear wheels, and 108mm width on the front, giving a total braking area of $0.74m^2$.

The hand (parking) brake operated via a cross shaft acting mechanically on two friction pads bearing on a disc fitted to the rear of the auxiliary gearbox. Alternative designs were employed according to the vehicle role. A conventional ratchet-operated lever was provided in the cab.

Road wheels
Wheels were 10.00x20in four-piece disc type, with a separate locking rim, mounting 15.00x20 directional or simple bar-grip type cross-country tyres and tubes. Non-skid chains could be fitted to the front wheels, and overall chains to the rear.

The recovery vehicles carried a single spare wheel, carried flat on the rear deck beneath the crane jib. The FV1103 artillery tractor carried two spare wheels, one each inside stowage compartments at the rear of the body on either side; a detachable jib could be mounted to the tilt framework for handling the spare wheel, and a hinged compartment door doubled as a runway in the open position. On the FV1122 tractor, the chain hoist provided for raising the gun trail could also be used for handling the spare wheel.

The braking system compressor could also be used for tyre inflation.

Chassis
Deep channel-section side members formed the main chassis rails, with six or seven channel-section or fabricated cross members, according to the length of the chassis, and two tubular cross members. Located behind the front-spring cross member on both chassis types was a fabricated cruciform bracket designed to support the ball end of the front axle torque frame. Cast-steel brackets at the rear, with pressed-in bronze bushes, supported the rear spring eyes.

A towing pintle was attached to the rear cross member. The front bumper was bolted directly to the underside of the main channels, and was also used to mount a towing hitch.

The chassis frame used on the recovery vehicle was a 'composite' version of the tractor and cargo truck frames. The rear cross member and tie channels were omitted, and the frame members modified to accept the subframe for the recovery equipment. The front bumper included a vice mounting, while the rear crossmember was modified to mount hydraulic outrigger jacks.

Cab and bodywork
Although the basic cab and front-end sheet metal was essentially the same on all the variants, the cabs differed in detail, one to another. A short, deep cab was used on the FV1119 recovery vehicles, with a longer, crew cab on the FV1103 artillery tractor, which was also fitted with armoured floor plates. In addition, each vehicle had a unique rear body according to its designated role.

The headlights were mounted either above, in a choice of two positions, or below the front bumper.

Cab
The steel cab consisted of a welded-steel frame with steel outer panels riveted to it; the outer panels, which were of double-skin construction, had 'Isoflex' insulation incorporated between the inner and outer skins. The cab consisted of upper and lower assemblies, bolted together in such a way that they could easily be dismounted to reduce the overall height for shipping.

A removable, hinged two-part hatch was provided in the cab roof above the passenger's seat, with the hatch halves hinged to the sides, and with each half able to be opened separately. There was provision for mounting a light-machine gun over the hatch, where the roof was specially strengthened.

The FV1103 artillery tractor was fitted with four hinged doors, two each side; all other versions had just one door each side. Steel ladder type steps were provided beneath each door. The front doors in each case were key-lockable. All doors were fitted with drop-down windows, with a fixed quarterlight installed between the door and windscreen pillar. The crew cabs were also fitted with sliding windows behind the rear doors. The split windscreen consisted of two fixed lower lights, together with two top-hinged upper lights. The original short cab, used on the cargo and FV1119 vehicles, also included two small glazed, sliding rear screens behind the driver and passenger.

The short cab used on the FV1110 and FV1119 vehicles extended some 300-400mm below the doors and was provided with stowage lockers in this space, where the crew cab had open racking below the doors.

At the front, the cab was bolted to a steel dash plate and bonnet support assembly, and attached to the chassis. The support assembly also provided a mounting for the wooden cab floorboards.

The bonnet, wings, and radiator grille were separate. The bonnet consisted of a fixed centre section, incorporating a hinged centre inspection panel; this was attached to the bonnet support and radiator. Two double-hinged opening panels provided access to the engine at each side.

Early vehicles used 'cycle' wings, but on the production examples, massive pressed-steel wings were bolted to the bonnet support panel and chassis side members, with treadgrip strips attached to the tops of the wings. The radiator was provided with side panels and a steel front guard assembly.

The smaller cabs were designed to accommodate a crew of two or three. The driver was provided with an upholstered bucket seat, adjustable for height and reach; there was a fixed bench seat with separate back rests for up to two passengers. A platform was provided beneath the hip ring for use by an anti-aircraft gunner. The crew cab included two additional rows of bench seating for nine passengers.

Armour plates were incorporated below the floor and toeboards, and behind the rear seats on the FV1103 artillery vehicle, sufficient to provide protection against a 4lb mine exploded by either the front wheels, or the leading wheels of the rear bogie.

FV1103 artillery tractor bodywork
The rear bodywork of the artillery tractor was constructed from wooden sections on a steel framework. Although the tailgate was hinged, the sides were fixed. A pressed-

steel mudguard assembly covered both rear wheels; the prototype vehicle had shorter panel behind the rear wheel, with a separate mudflap.

Like the cargo versions, a removable canvas cover was supported on four steel hoops to cover the cargo area.

Recovery equipment
The recovery equipment was mounted on a subframe of supporting channels, to which was mounted a simple steel platform body. Stowage lockers in the platform were provided for the various items of loose equipment. Stabiliser brackets and outrigger jacks were attached to the subframe at strategic points to assist in recovery operations, and a jib stay and 'A' frame bracket was bolted to the rear support channels; the 'A' frame could be fitted in one of two positions. A hinged hydraulically-operated spade-type earth anchor at the rear of the vehicle allowed pulls up to 30 tons.

A dual-section extensible jib, mounted on a vertically-pivoted post, was bolted to the recovery equipment subframe. Jib extension, luffing and slewing actions were provided by means of hydraulic rams. Hoisting was effected by a hydraulic winch mounted at the end of the jib, with a disc brake to hold the load. Provision was made for single- or twin-fall reeving of the winch rope by twin pulleys on the jib; a twin sheave block was carried for 4-fall reeving.

Jib control was carried out from a cab on the right-hand side of the outer column. There was a protective cage fitted over the cab on prototype vehicles only, although this item was also supplied loose on production examples.

Winch
Two types of winch were employed according to role; a mechanical winch was fitted to the FV1103 artillery tractor, and a hydraulic winch was installed on recovery vehicles.

Mechanical winch
The artillery tractors, were fitted with a mechanical 8 or 10 ton, vertical drum winch, manufactured by Wilde, and driven from a power take-off on the auxiliary gearbox. The winch frame was bolted to the rear end of the chassis, with fairleads at front and rear and cable rollers installed along the chassis. Average rope speed, at an engine speed of 1000rpm, was 4.57, 7.62 and 3.66m/min, in first, second and reverse gears respectively.

Engagement of the winch was effected by means of a sliding dog in the auxiliary gearbox, controlled by a lever in the cab. An electrical cut-out circuit, wired through the ignition, was provided to protect the winch against overload.

Hydraulic winch
Recovery vehicles were fitted with a two-speed, 15 ton, horizontal-drum hydraulic winch installed under the rear of the cab. Rope fairleads were provided at front and rear, and cable rollers were installed along the chassis. Average rope speed, for a 15 ton pull, in low gear, was 4.57m/min, and for a 5 ton pull, in high gear, 13.73m/min.

The winch was powered by a reversible hydraulic motor, through a two-speed epicyclic gearbox; the motor itself was driven from a power take-off on the auxiliary gearbox. A pneumatic rope tensioner was fitted. Engagement of the drive for the winch motor was effected by means of a sliding dog in the auxiliary gearbox, controlled by a lever in the cab. The winch was protected against overloading by a pressure-relief valve in the hydraulic circuit.

Electrical equipment
The standard 24V negative-earth electrical system was employed, using FVRDE-designed components throughout. Power was derived from a pair of 12V 60Ah batteries, with the charge maintained by a 'No 1, Mk 2', or 'No 1, Mk 2/1' generator. The maximum generator output was 12A at 1260-1410 or 1100rpm (generator speed) for the two types respectively.

The starter was a 'No 1, Mk 2' or 'No 1, Mk 2/1' machine, rated at 3.5 or 2.5hp respectively.

DOCUMENTATION
Technical publications
Specifications
War Office Specification 70/Vehs/81 GS(W)2: tractor, 6x6, medium artillery, FV1103.
FVRDE Specification 9083. Production specification: body and mounting for recovery vehicle, wheeled, heavy, 6x6, CT, Leyland, FV1119.
FVRDE Specification 9171. Production specification: chassis for recovery vehicle, wheeled, heavy, 6x6, CT, Leyland, FV1119.

Reports
BAOR Report N5870 Tech. Final report on tractor, 10 ton, 6x6, CT, medium artillery, Leyland. 6 January 1956.
FVRDE Report FT/B 911. Truck, 10 ton, 6x6, FV1121, towing 8in howitzer.
Australian Army Design Establishment Report TI1919, Parts 1, 2 and 3. Tropical trials in Northern and Central Australia 1960/61.

User handbooks
Tractor, 10 ton, CT, med arty, 6x6, Leyland Mk 1. Army code 12239.
Recovery vehicle, heavy, 6x6, Leyland. Army code 13621.

Servicing schedules
Tractor, 10 ton, CT, med arty, 6x6, Leyland, Mk 1. Army code 13615.
Recovery vehicle, heavy, 6x6, Leyland. Army code 13608.

Parts lists
Recovery vehicle, heavy, 6x6, Leyland. WO code 13837.
Truck, cargo, 10 ton, 6x6, Leyland; tractor, wheeled, GS, 10 ton, 6x6, Leyland. Army code 14886.

Technical handbooks
Data summary:
Truck, 10 ton, GS, cargo, 6x6, Leyland Mk 1; tractor, 10 ton, GS, med arty, 6x6, Leyland Mk 1. EMER N620.
Recovery vehicle, heavy, 6x6, Leyland. EMER N620/1.

Technical description:
Truck, 10 ton, GS, cargo, 6x6, Leyland Mk 1; tractor, 10 ton, GS, med arty, 6x6, Leyland Mk 1. EMER N622.
Recovery vehicle, heavy, 6x6, Leyland. EMER N622/2.

Repair manuals
Unit repairs:
Tractor, 16 ton, med arty, 6x6, Mk 1, Leyland. EMER N623.
Recovery vehicle, heavy, 6x6, Leyland. EMER N623/1.

Field repairs:
Tractor, 16 ton, med arty, 6x6, Mk 1, Leyland. EMER N624.
Recovery vehicle, heavy, 6x6, Leyland. EMER N624/1.
Base repairs:
Tractor, 16 ton, med arty, 6x6, Mk 1, Leyland. EMER N624, Part 2.
Recovery vehicle, heavy, 6x6, Leyland. EMER N624/1, Part 2.

Complete equipment schedules
Tractor, wheeled, GS, med arty, 10 ton, 6x6, Leyland. Army code 33832.
Recovery vehicle, heavy, 6x6, Leyland. Army codes 30552/1, 33833.

Bibliography

A war horse in mufti. London, Commercial Motor magazine, 13 August 1954.

Breakdown. A history of recovery vehicles in the British Army. Baxter, Brian S. London, HMSO, 1989. ISBN 0-112904-56-4.

Leyland Martian civilian model 6x6 tractor. London, Commercial Motor magazine, 24 September 1954.

The Leyland Martian. Profile of the British Army's FV1100-series 10 ton 6x6 range. London, Wheels & Tracks, no 23, 1988. ISSN 0263-7081.

Above. A pair of well-restored O853 AEC 'Matadors'. The vehicle nearest the camera has a gas detector plate beneath the windscreen, while on the further vehicle, half of the canvas cab roof has apparently been replaced by metal. Despite its WW2 serial number, the second vehicle is painted in post-war Deep Bronze Green.

Left. Another O853 'Matador', this time in overall green finish, with a metal top to the rear body.

Top. Nicely-restored O853 'Matador' with WW2 'Mickey Mouse' ears camouflage, parked in Bethune town square, as part of the liberation celebrations, 1995. The AEC triangle on the radiator is not original, but the vehicle was towing a 25 lb field gun, which was a nice touch.

Top left. Surplus Matadors were very popular with forestry contractors and showmen. This one, which leaves no doubt as to its make and model, is operated by Harris's Amusements. The exact purpose of the square glazed lights above the windscreen is unclear.

TUGS OF WAR

Left. **Mk 3 'Militant' recovery tractor (FV11044)** from the REME/SEME historical vehicles collection, based at Bordon in Hampshire.

Left. **FV11002 6x6 Mk 1 'Militant' artillery tractor.** The front tow hook is not original, and the winch fairlead rollers can just be seen on the bumper to the extreme right.

Above. **A pair of Mk 3 'Militant' cargo trucks**, one of which carries a Bedford MK. Note that although it is similar to the recovery tractor, the cab is shorter front to back. Both vehicles have a ventilation inlet beneath the driver's windscreen.

Above. **Close-up view of the Mk 3 'Militant' cab, once again on a cargo vehicle. Compare this to the recovery tractor on p78 and note the lack of pusher pads on the front bumper.**

Top right. **Albion CX22S artillery tractor shows off its unusual radiator configuration.**

Right. **Helen - a nicely-restored Albion CX22S artillery tractor. Note the radiator muff and extension to the overflow pipe, neither of which are present on the vehicle above.**

Above. A coating of Beltring dust adds a nice touch of authenticity to this pair of Diamond T ballast tractors. On the left, the closed cab version has its windscreen cracked open a touch; on the right, the open cab equivalent, provides considerably more ventilation, with rather less mechanism. The owner of the open-cab version has christened the truck, and identified the engine, with admirable economy.

Left. Closed-cab Diamond T 981. Note the canvas enclosure to the ballast box, and the head and side lamps in the position favoured by the British Army on rebuilt post-war examples.

TUGS OF WAR

Above left. **I make no excuses for including this shot of the remains of a Diamond T wrecker.** While the bonnet, bumper and front wings are different on the 980/981, the open cab is shared with the ballast tractor.

Above right. **Diamond T and obviously proud of it!** This closed-cab ballast tractor seems to carry the lights from a WW2 half-track, but the photograph shows well the brake-line air coupling on the front valance. The cab is equipped with the AA hip ring, but there seems to be no roof-level vent.

Right. **Nicely-restored and very usable 980 closed-cab ballast tractor.** While the massive rear-view mirrors might not be original, they certainly make it easier to thread the beast through modern traffic.

A nice contrast between the art deco styling of the closed cab Diamond T, and the austere perpendicular lines of the open cab. Note the difference in lamp positions on this pair of 981 ballast tractors.

FV1119 'Martian' recovery tractor. This very authentic-looking vehicle belongs to the REME/SEME historical vehicles collection. Not all of the 'Martians' showed the maker's name on the bonnet sides.

TUGS OF WAR

Whilst the thirst of the Rolls-Royce B81 engine did not endear the 'Martian' to commercial recovery operators, its prodigious pulling power, combined with low price, meant that, from time to time, one might encounter an FV1119 recovery tractor parked behind a country garage, for the odd heavy recovery job (right). However, with a total production of 280 vehicles, most 'Martian' recovery tractors were destined for the blast furnace (above). Here, a clutch of recovery tractors are slowly overtaken by nature.

Above left. £22,500 worth of quality military hardware slowly returns to its constituent parts. An FV1119 recovery tractor disintegrates in a Dutch surplus yard.

Above right. Well-restored 'Martian' recovery tractor shows off the jib, crane operator's cab, spade anchor, 'A-frame' and rear winching arrangements.

Left. Towards the end of the 'eighties the two FV1200 tractors were broken up at Hardwick's yard in Ewell, Surrey. Here, prototype number 2, without its engine, but otherwise very complete, awaits the cutting torch. Note the civilian registration and FVRDE markings. Photo by Mike Maslin.

Top. **R100 'Pioneer'** artillery tractor. Note the anti-reflective netting rolled-up above the windscreen, and the spare wheel stowed on the roof of the rear body.

Above. At some time during its chequered career, the second FV1200 prototype acquired twin rear wheels and these massive wheel arch extensions. Photo by Mike Maslin.

Right. 'Pioneer' recovery tractor shows the austere lines of the cab.

Left. **Sweet Daisy**, a nice 'Pioneer' SV/2S recovery tractor in front of Duxford's Hall of Land Warfare. The weights above the front axle help to counter the weight of a recovered vehicle.

Left. Close-up view of the 'Pioneer' engine room. Note the external air cleaner, and the water gauge attached to the top of the 'coffee pot'. The slotted plates forward of the radiator provide three positions for the pig-iron counterweights.

Above. Front view of an R100 artillery tractor, shows the stark simplicity of the bodywork of the 'Pioneer'. Some examples carried a cast nameplate on the radiator plinth.

Above. 'In case of doubt - brew up'. Big John finds a novel use for the front tow hook.

Top right. When 'Pioneers' were common on the surplus market, their slow speed and low gearing endeared them to road makers. Shown here in the colours of W & J Glossop, this 'Pioneer' TRMU/30 fifth-wheel tractor once towed a heater/planer for removing road surfaces.

Centre right. 'Pioneer' SV2/S recovery tractor in RAF livery. The cage beneath the cab was used for carrying the tracks for the rear bogie, as well as other recovery kit.

Right. An 'Explorer' recovery tractor demonstrates the general similarity with the pre-war 'Pioneer'. The photo shows the spare-wheel mounting, additional rear stowage boxes, and the 'A-frame' towing device. The jib is in the stowed position.

TUGS OF WAR

Left. 'Constructors' are not common, either among collectors or with commercial operators. Here, in the livery of Frodsham Motors, a well-loved FV12102 fifth-wheel tractor has been converted to a recovery vehicle. The cab roof cover has been removed and the vehicle carries various additional guards, but the Rolls-Royce engine remains in place and most of the military fittings are present and correct.

Left. Another view of Frodsham's FV12102-based recovery tractor shows the squared wings, which with the narrow cab, and below-door fuel tanks conspire to give an American look to the truck.

Above. A trailer ramp provides a makeshift articulation gauge, allowing this FV11301 'Explorer' to demonstrate the flexibility of its chassis and suspension.

88   TUGS OF WAR

Above. Close-up view of this re-engined and well-detailed 'Explorer' recovery tractor, shows the general similarity with the pre-war 'Pioneer'. The air-conditioning unit on the cab roof is a modern addition, which probably indicates that this truck still works for a living. The markings on the wing are those of REME.

Top right. Alongside the 'Matador' tractor shown on p77, Harris's Amusements also operate this 'Explorer'. The air cleaner seems to have been changed to a commercial type, and the badge on the side of the radiator indicates that the thirsty Scammell-Meadows engine has been replaced by a more-economical Rolls-Royce diesel.

Right. 'Explorer' recovery tractor with the jib in the fully-extended position.

TUGS OF WAR

89

Good side elevation of FV11301 'Explorer' showing jib and winch, and the bottom-hinged door and ladder which provide access to the crane operator's position. This vehicle does not have the additional stowage boxes shown on p87. Note also the distinctive Scammell 'coffee pot' on the radiator.

Kicking up the dust - an 'Explorer' recovery tractor shows that not all restored vehicles are confined to static display. Note the precautionary recovery cable wrapped around the radiator guard.

FV121001 Mk 1 'Antar' with steel ballast body. With a total width of 2.82m, the caution board was a necessary fitment for road use. This is another vehicle from the REME/SEME collection.

Right. FV12004 Mk 3 'Antar' fifth-wheel tractor. The re-styled cab and front-end sheet metal provide a less massive appearance, and the use of just one radiator permitted the grille to be narrower. Nevertheless, the Mk 3 'Antar' is still sufficiently large to require the same width warning as on the Mk 1. Note the massive towing pin and 'A-frame' brackets on the front bumper.

Above. FV12003 Mk 2 'Antar' with wooden ballast body. The concrete ballast weights can just be seen behind the cab. The framework above the cab roof is an anti-aircraft gun mount.

Right. Close-up view of the winch operator's position on the REME Mk 1 'Antar'. To the right of the photograph can be seen the twin fuel tanks.

Imagine seeing this in your rear-view mirror - at least you'd not be left wondering what it was! The military 'Antars' never were officially dubbed 'Mighty', though this had no effect on their physical size. Note the twin radiator fillers.

American-built Rogers M9 trailer hitched to a Diamond T 981 ballast tractor; the complete tractor-and-trailer rig was known as 'tank transporter M19'. These 45 ton trailers were the largest available for tank transporting during WW2.

CHAPTER 4.3

LEYLAND FV1200 SERIES
FV1201

Alongside the FV1000 60 ton super-heavy tractor, the Ministry of Supply plan for post-war CT vehicles had also included a requirement for a 30 ton heavy tractor. It was intended that this would replace the ageing WW2 Scammell 'Pioneer' artillery tractors and tank transporters, and the American Diamond T 980/981 ballast tractors, but the Ministry also had a range of other roles in mind.

Designated FV1200, the first vehicle to be produced in this series was a heavy artillery tractor (FV1201), but the range was also intended to include a tractor for the 50/60 ton tank-transporter trailer (FV1202), road-going and cross-country tractors for 30 and 60 ton semi-trailers (FV1203, FV1204, FV1206, and FV1207), including one variant possibly designed for use with a powered semi-trailer, and a heavy recovery vehicle (FV1205).

The original design work was carried out by Dennis Brothers of Guildford, and it was also intended that they would produce the prototypes.

An announcement had appeared in Commercial Motor magazine in 1949 stating that Dennis Brothers were involved in 'important' work for the Ministry of Supply, and it seems that the project finally got underway during 1950. However, problems described at the time as 'material shortages and changes of personnel', meant that Dennis were unable to see the contract through to completion. Worse still, contemporary FVRDE records suggest that Dennis were not even able to complete the first and second stages of the contract. This had called for the preparation of a full-size mock-up of the front end of the vehicle, together with a cooling rig intended to be used in conjunction with the prototype development work.

Although in 1952, Dennis had shown models of their proposed FV1201 and FV1206 prototypes at FVRDE, in 1953 the contract was transferred to Leyland Motors, with the work being undertaken at the Leyland Ministry of Supply factory in Lancashire. Eventually two prototypes of the artillery tractor were produced by Leyland and made available for testing during 1954. Having completed construction of the prototypes, Leyland estimated that the cost of the production vehicles would be some £20,000-22,000 each, plus £250,000 for tooling, of which £100,000 was to be made as a grant by the Ministry of Supply towards the cost of machine tools.

In 1955, the two prototypes were put through their paces at Chertsey and elsewhere, but once again, the project

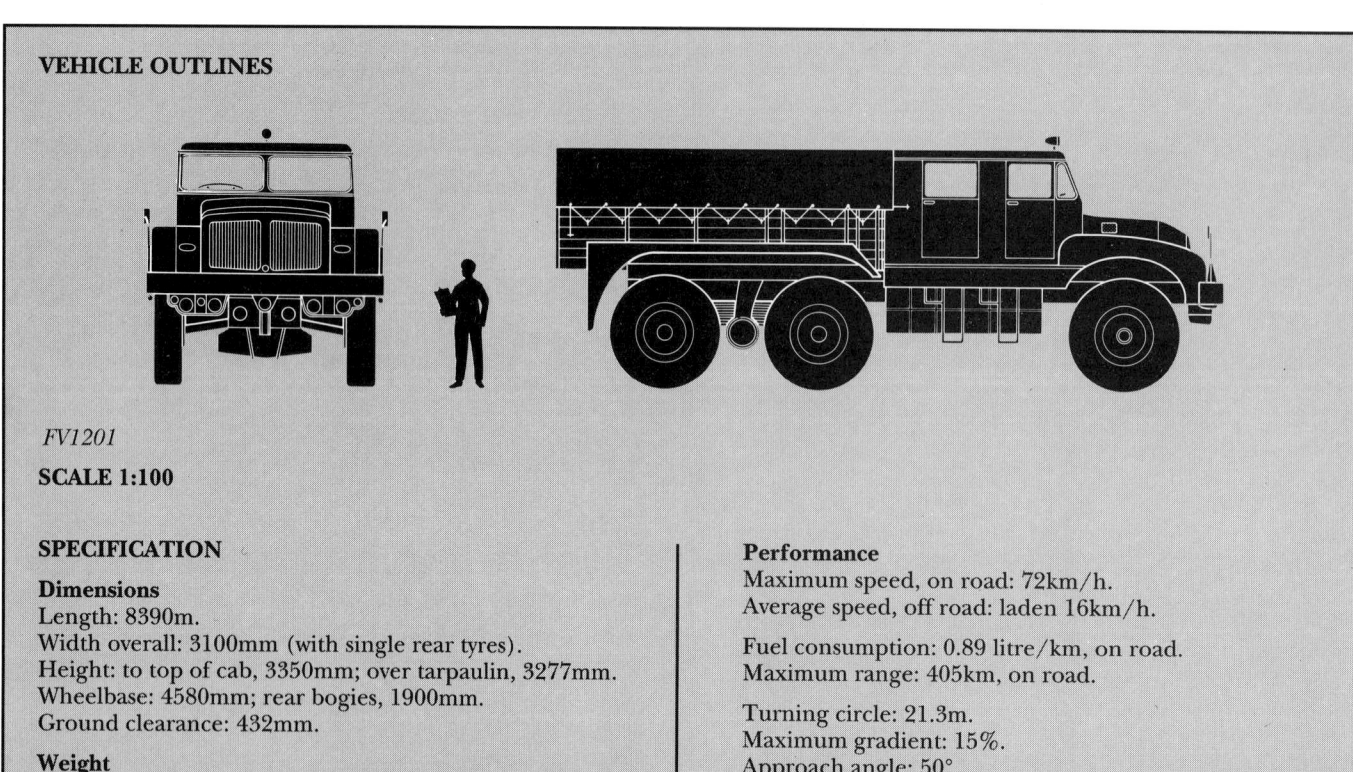

CHAPTER 4.3: LEYLAND FV1200 SERIES

VEHICLE OUTLINES

FV1201

SCALE 1:100

SPECIFICATION

Dimensions
Length: 8390m.
Width overall: 3100mm (with single rear tyres).
Height: to top of cab, 3350mm; over tarpaulin, 3277mm.
Wheelbase: 4580mm; rear bogies, 1900mm.
Ground clearance: 432mm.

Weight
Laden: 33,636kg.
Unladen: 24,945kg.

Bridge classification: not classified.

Performance
Maximum speed, on road: 72km/h.
Average speed, off road: laden 16km/h.

Fuel consumption: 0.89 litre/km, on road.
Maximum range: 405km, on road.

Turning circle: 21.3m.
Maximum gradient: 15%.
Approach angle: 50°.
Departure angle: 40-50°.

Fording: unprepared, 762mm; prepared, 1980mm.

Capacity
Maximum towed load: 30,000kg.

---

was destined to come to nothing, and even while the testing was underway, the range was simplified to include only the FV1201 and FV1205 variants.

In the end, the FV1201 artillery tractor prototypes were the only examples to be constructed, and none were put into quantity production. Within a year, the artillery tractor role had been cancelled and replaced by the 10 ton 'Martian' (see Chapter 4.2), while the Thornycroft 'Antar' (see Chapter 6.1) was purchased as a 30 ton tractor for use with full and semi-trailers. None of the other roles was pursued.

DEVELOPMENT

The FV1200 range was intended for towing heavy artillery such as the Vickers 4in gun, for semi-trailers carrying AFV's, for recovery trailers being used to recover disabled AFV's, for use as a tank tractor, and as a breakdown recovery vehicle.

In an early draft of the plan for post-war vehicles published in 1945, the FV1200 range was stated as being '10 tons, long chassis', with the same plan describing the FV1100 range (see Chapter 4.2) as '10 tons, short chassis'. By 1949, the weight classification had risen to 20 tons and the vehicle was being described as the 'medium tractor'. It is not clear whether these weight references were errors but certainly by the end of 1949 the weight had finally been upgraded to 30 tons.

The design brief for the FV1200 range originally described a vehicle with alternative wheelbase lengths and a choice of tyre and wheel equipment to cater for different roles. The abandonment of the FV1000 tractor due to problems with cost and weight, led to the inclusion of the 60 ton off-road transporter role to the list of tasks assigned to the FV1200, and this resulted in the design of a second rear bogie configuration. The two types of rear axle equipment planned were a walking-beam type bogie designed to accept single or twin tyres, and a conventional tandem axle intended only for twin tyres.

Although Dennis had demonstrated both axle configurations with scale models in 1952, and both single and twin rear wheel versions were tested, only chassis incorporating the first bogie design were constructed.

Outline specifications and simple schematic drawings were prepared by FVRDE for most of the 30 ton FV1200 Series variants, and were published for discussion in 1949, as well as appearing in a so-called information brochure dated late 1951 which seemed to suggest that work was well in hand. In October 1952, a meeting was convened at FVRDE with a view to discussing and finalising the users' requirements for the complete range. However, none of the proposed vehicles was to make it into production, and in practice only the FV1201 heavy artillery tractor variant was even prototyped.

A development contract (6/Veh/5747/CB27a) for the FV1200 Series had been placed with Dennis Brothers, which included the construction of six prototype vehicles. Two of these were to be of the FV1201 artillery tractor; two FV1205 recovery tractors, intended to replace existing armoured recovery vehicles (ARV's); one fifth-wheel tractor for cross-country use (FV1206); and one for road use (FV1207).

Meetings were held between Dennis Brothers, FVRDE, the Ministry of Supply, and representatives of the user arms at Chertsey in October 1952, where the basic design specification was discussed, and where Dennis personnel showed drawings and scale models of the FV1201 and FV1206 tractors. Following these meetings, the War Office suggested to FVRDE that both the artillery tractor and the cross-country fifth-wheel tractor intended for use with the proposed FV3101 60 ton semi-trailer should be raised to priority '1'. The reason given was that at that time, there was no suitable artillery tractor available, and it was planned that deliveries of the Vickers medium AA gun would begin in 1955, with the ADE 'Red Maid' medium AA gun due for production in 1958, and the new 175mm field gun scheduled to appear in 1957. It was felt that prototypes of the vehicles should be available at the same time as the guns.

Although, in the end, there was no progress made with the development of the FV1204 fifth-wheel tractor, it is interesting to note that the specification which had been issued in August 1949 had stated that 'consideration should be given to taking the drive to the semi-trailer wheels', this was presumably in an effort to improve cross-country performance, and pre-dates the powered Scottorn trailers used with Land Rover 101 vehicles by some 25 or 30 years.

However, all was not well with the development programme. Dennis had many years' experience of truck chassis, but they had tended to concentrate on lighter vehicles, and perhaps the 30 ton FV1200 Series proved to be beyond their capabilities. For whatever reason, the work was not completed, and in 1953, the Ministry of Supply transferred to project to Leyland Motors, giving them a contract for the completion of the work begun by Dennis, together with the preparation of general arrangement, unit assembly and detail drawings (the contract specifically stated that these should be 'in pencil... on paper'!), and the manufacture of the six prototype vehicles, together with spare parts as agreed with the project engineer.

As with the FV1000 project, motive power was to be provided by the Rover 'Meteorite' V8 petrol-injection engine, which was effectively two-thirds of the successful V12 Rolls-Royce 'Meteor' and 'Merlin' engines which had powered both aircraft and tanks during WW2. Design work was carried out under the management of Messrs Tattersall, Thorpe and Wells at Leyland, with a Mr Kightley, who had previously been with Dennis Brothers, assigned to the work as project engineer by FVRDE - perhaps he was one of the mysterious 'changes of personnel' referred to in the earlier Ministry memo. Inspection of the prototypes was to be carried out, on behalf of FVRDE, by W G Whitehead.

The engine cooling and test rig, which had formed the first stage of the original contract, was completed by January 1954, while the manufacture and machining operations for construction of the first vehicle were begun in November 1953. The first prototype vehicle was completed and despatched to Chertsey in July 1954, albeit without its body (this was not available until late August). The second vehicle was delivered in October 1954, but in December of that same year, the contract was amended to cover just the two prototypes and numbers three to six were never constructed.

Considerable difficulties were encountered during the machining operations with poor-quality castings which often proved either to be porous and thus completely unsuited to their purpose, or so short of material on faces which were to be machined, that remedial work was required before machining could begin. It may well be that this was a common problem during these immediate post-war years as the automotive and engineering industries struggled to return to normal production with worn-out machinery and materials shortages. There were also problems with the dimensions of components produced by sub-contractors and a number of modifications were made to accommodate these deficiencies.

On 7 July 1954, while still at Leyland's works, for the first time the prototype vehicle travelled under its own power. At the dizzy speed of 25km/h, the gearbox seized solid due to inadequate working clearances. The problems were remedied by stripping and remachining, and a 15km track test suggested that all was now well. Difficulties were also encountered with the hydraulic steering pump;

CHAPTER 4.3: LEYLAND FV1200 SERIES

these were dealt with by a representative of Lockheed, before delivery of the prototype to Chertsey.

The problems experienced with the first prototype led to some rethinking, and a number of detail changes. Prototype number two was fitted with a modified design of walking beam, and incorporated a power take-off and winch-drive gearbox mounted behind the main gearbox. Before trials commenced, modifications were also made to the gearbox of the second vehicle to prevent it jumping out of gear.

The trials also highlighted difficulties with the drive to the walking beams and new parts were designed and produced, before being despatched to FVRDE for incorporation into both prototypes.

At the end of the 1955 trials, it was suggested that most of the detail points which had arisen could be easily dealt with; and in fact, many of them had already been attended to during the trials phase. The only points given specific mention in the inspector's report were relatively minor: for example, it was felt that the side-mounted fuel tanks were particularly vulnerable to damage from the crew's feet as they climbed into the vehicle; the second point concerned the driver's seat, which it was felt did not offer sufficient side support during off-road operation.

Although the FV1200 Series was more-or-less approved by FVRDE, this was not a view shared by the policy makers at the War Office who felt, that at the quoted price of £22,700 each for a quantity of 200, or £22,200 if 500 vehicles were ordered, these tractors were way too expensive. In 1949, the cost had been estimated at just £11,000 each, and to put these figures into context, at the time, the cost of the 'Antar' was £12,000 and the Scammell 'Constructor' just £7350.

It had been suggested by the user arms that some 560 FV1200 tractors would be required, 400 for use as artillery tractors, and 160 recovery vehicles. A strong case for placing the production contract was being made by the Director, Mechanical Engineering (DME) and the Director, Weapons Development (DWD) on the basis that operational advantages outweighed the potential cost and maintenance difficulties. They were obviously not convincing, and in 1955, W A Scott, Director, Weapons Development finally admitted that the 'War Office cannot afford this vehicle'. It was proposed that the Royal Artillery should accept the slightly inferior performance of the Scammell 'Constructor' (see Chapter 5.2), while REME should 'seek a cheaper alternative to the FV1200'.

It is interesting to note that long before any trials had begun, Thornycroft had already introduced the 'Antar' which, in certain roles, seemed to offer comparable

Artist's impression of FV1201 produced by Dennis Bros (IWM)

Prototype under construction by Leyland (BCVM)

Good interior view showing control column and brake compressor (BCVM)

96   TUGS OF WAR

First prototype on delivery to FVRDE (IWM)

Side elevation showing temporary rear wings and lack of bodywork (IWM)

... from some angles, the prototype was almost stylish (IWM)

(road) performance at a more reasonable cost, and even as far back as 1949 the War Office was investigating the possibility of using 'Antars'. Although the first vehicles entered service in 1951, the 'Antar' was not really being seen at the time as a replacement to the FV1200, more as a stop-gap measure. It had even been stated in December 1954 that 'we should not interfere with the plot for increased production of 'Antars' because of the possibility of using FV1205... if it (FV1205) is brought into service, our orders for 'Antars' could be cut down in later years'.

Of course, the 'Antar' was ultimately to emerge as an extremely capable vehicle which was to remain in service until well into the 1980's. However, the 'Antar' had no significant off-road capability and there were situations where it would be necessary to recover and transport disabled tanks from possibly extremely-rugged terrain, and it was still believed that the FV1200 might have a role to play.

As early as 1954, it had been suggested that the FV1200 range be expanded to include a wheeled tank recovery tractor, intended for dragging disabled AFV's to a suitable location for more-conventional recovery. In this role, the FV1200 would bridge the gap which existed between the ARV's with their limited road-going characteristics, and the wheeled tank transporters which had very limited cross-country abilities.

By 1956, the prototype had been turned over to REME and it was felt that the FV1201 might be able to serve as a tank tractor, removing disabled AFV's from forward areas, a role normally reserved for the tracked armoured recovery vehicles (ARV's). The vehicle was trialled in this role, and photographs exist of one of the prototypes towing a 'Conqueror' tank, possibly in Germany.

So, the War Office did not discard the FV1200 quite as lightly as the FV1000 had been abandoned, and the testing did not end with the Chertsey trials, but actually moved into the user trials phase. It was not until February 1960 that it was finally admitted that there was no future for the FV1200 and the testing phase was officially concluded. However, since the tooling had been disposed of in February 1959, it is hard to understand why the tests were not terminated at the same time.

These were massive vehicles, with a width way over that normally seen on British roads. Looking back, it is hard to understand why such an approach was adopted, and one should not under-estimate the difficulties of piloting a vehicle, which was close to 50% wider than the then-current legal maximum, through the narrow streets of the average British or German small town. Perhaps this had also contributed to the general unhappiness with the project.

CHAPTER 4.3: LEYLAND FV1200 SERIES

Anyway, by 1959 when the tooling was scrapped, the writing was clearly on the wall for the FV1200 and, when in 1960 it was finally admitted 'that there was no need for this tractor', that was the end of Leyland's second foray into the heavy tractor market. Like the FV1000 project, the FV1200 Series was abandoned. It's ironic that although Chertsey rarely photographed one vehicle against another, contemporary photographs exist of these two giants, side by side - Gog and Magog.

The two FV1200 prototypes remained at FVRDE for a while, eventually ending up in Hardwick's yard at Ewell where they remained until at least 1988, but ultimately they were both cut-up for scrap. By the time the trucks had found their way to Hardwick's, at least one of them had been fitted with twin rear wheels with 'Rock Grip' type tyres, and the rear mudguards had been extended to cover the increased width, giving a most bizarre rear aspect.

It is said that the walking-beam bogies were sold complete, to a shipyard where they were mounted sideways, still wearing their massive 18.00x24 cross-country tyres, on either side of a dry dock to help guide the ships into position. But what a pity neither vehicle was saved!

NOMENCLATURE

FV1201. Artillery tractor, 30 ton, CT, heavy, 6x6, Leyland.

There were no variations: although other variants were specified, and details appeared in print, none were produced.

DESCRIPTION

Engine

As with the larger FV1000, it was originally intended that the vehicle be fitted with a diesel engine, but in the event, the Rover 'Meteorite' V8 petrol engine was used as a stop-gap measure.

At a time when the standard military fuel was 70 octane rating petrol, the Mk 1A 'Meteorite' was designed for use with minimum 80 octane rating fuel and the War Office seemed prepared to accept the resulting logistic problems of having to handle two grades of fuel, in return for the increased power.

The version of the 'Meteorite' installed in the FV1200 employed overhead inlet and exhaust valves, with four valves and two spark plugs per cylinder, together with a mechanical direct-injection system for fuel supply to the engine. The injection equipment, supplied by the SU Company, comprised an SUX 727 eight-point injector pump supplied by a pair of David 'Korrect' fuel pumps mounted on the block and driven by a cam on the

The tiny rear lights help to give some idea of the massive scale (IWM)

Rear view shows huge walking beams and brake drums (IWM)

Second prototype was delivered with body fitted (BCVM)

... and differed in small details (BCVM)

Offside view of second prototype (BCVM)

... and again from the rear (BCVM)

crankshaft. Air was fed to the engine via four **FVRDE** standard oil-bath air cleaners, two mounted on either side of the engine.

Unlike its FV1000 bigger brother, at least the FV1200 Series appeared to stand some chance of meeting the FVRDE specification for fuel consumption which had stated that 'the fuel capacity will give a range of not less than 250 miles'.

The oil cooler tubes for the engine lubrication system were mounted between the cooling fans and the radiator.

The ignition was of the magneto type, employing twin British Thompson Houston (BTH) Type C8B magnetos producing 12kV; the engine was perfectly capable of running on either magneto should one of them fail.

Engine data
Rover 'Meteorite 80' No 2 Mk 1A petrol injection, water-cooled V8; cylinder blocks arranged at 60° angle.
Capacity: 18,012cc.
Bore and stroke: 5.40 x 6.0in.
Compression ratio: 7:1.
Firing order: (offside cylinder block designated 'A', nearside block 'B') A4 B1 A2 B3 A1 B4 A3 B2.
Power output: 510bhp (gross); max engine speed 3000rpm; governed speed 2800rpm.
Maximum torque: 1090 lbs/ft (gross) at 2200rpm.

Cooling system
The engine was water cooled, with the cooling system pressurised to 7050kg/m². The cooling water was passed through a pair of Morris 'H' matrix radiators with 10 rows of tubes, giving an overall thickness of 125mm. A massive 10-bladed cooling fan of some 550mm diameter was installed in front of each radiator, driven by triple belts.

Transmission
Like the FV1000 Series, drive was transmitted by means of a 560mm Hydraulic Couplings 'Fluidrive' viscous coupling, with a separate 405mm diameter break-away friction clutch. The main gearbox, which was manufactured by Dennis Brothers, provided five forward and one reverse speed, and was fitted with a two-pedal gear change system with air servo-assistance. All transmission controls were housed in a cast-aluminium control tower positioned to the right of the driver; this fairly bristled with a total of six levers.

From the main gearbox, the power was transmitted to a unit-constructed three-speed auxiliary, or transfer gearbox, also produced by Dennis, and then by means of open propeller shafts running the length of the vehicle to drive the front and rear axles. The transfer box ratios were 1:1, 1.37:1, and 2.74:1. The auxiliary gearbox was

CHAPTER 4.3: LEYLAND FV1200 SERIES

also provided with a power take-off intended to drive a mechanical winch, and with a nine-cylinder swash plate oil pump which was designed to drive the hydraulic crane of the FV1205 recovery version.

Drive to the rear axles was by means of bevel-and-spur reduction gears to the walking-beam centres, with a final chain drive to the wheels through the walking beams. The beam pivots were located on the same side of the beam as the stub axles, with the beam pivoting on a vertical arm attached to the chassis.

The solid front axle casing incorporated a bevel-gear type differential from which power was transmitted, via axle shafts and bevel gears, to the road wheels.

Suspension
A single transverse elliptical spring was installed across the chassis for the front axle, pivoted centrally in a swivel housing designed to allow plus/minus 305mm movement of the road wheels from the static position.

The original design called for half-elliptical rear springs, but as tested, the prototype FV1201 vehicle was unsprung at the rear; simple rubber buffers limited wheel movement to plus/minus 305mm from the normal position.

Steering gear
Steering was controlled through a Burman recirculating-ball type steering box with Lockheed hydraulic assistance. The steering action was transmitted to the wheel stations by means of an axle-mounted bell crank, rather in the style of the WW2 Jeep.

Braking system
All three axles were provided with twin-leading shoe, dual line air-operated brakes, actuated by a conventional foot pedal with servo assistance from piston-operated air cylinders.

The cast-steel brake drums were 483mm diameter x 241mm width on the rear wheels, and 137mm width on the front.

The hand (parking) brake operated on a drum mounted on the gearbox output shaft.

Road wheels
It was planned that both dual- and single-rear wheels would be used on FV1200 tractors, according to role; both wheel configurations were tested on the prototypes. On the FV1201, the wheels were 13.00x24in diameter with demountable rims: tyres were Goodyear 'All Service' NDCC or 'Rock Grip' type, 18.00x24. Smaller-section tyres, 13.00x24 mounted on 10.00x24in dual rear wheels, were planned to be used on the FV1204 tractors. The

Prototype FV1201 manages to make the Antar look quite modest (BCVM)

Is he trying to get in - or is he levitating? (BCVM)

In the mid 'fifties the first prototype went for trials in Germany (IWM)

CHAPTER 4.3: LEYLAND FV1200 SERIES

The first prototype was shown at the 1956 FVRDE exhibition (TMB)

Rear view of second prototype showing massive chassis (BCVM)

Both prototypes ended up at Hardwick's yard in the 'eighties (PG)

photographs of the two prototypes show that different tyres were fitted to each.

Non-skid chains could be fitted to the front wheels, with overall track type chains on the rear bogies, as were occasionally fitted to Scammells.

The FVRDE specification called for 'a spare wheel and tyre... (to) be provided in an easily accessible position'. The wheel was actually mounted in a recessed panel behind the nearside door; there was no sign of any mechanical handling device intended to assist with manoeuvring the wheel onto and off its mounting, but presumably there was some kind of davit intended to be fitted into the rear body.

Chassis

The chassis was of channel and box section design, with pressed and tubular cross members. The main chassis rails were designed to run more-or-less straight from front to rear without any sweeping-up over the axles; the rails were 100mm wide and 10mm thick, tapering from 225mm deep at the front, to 380mm deep at the rear. A full width, channel-section bumper was fitted at the front, with what appeared to be a massive spade-type earth anchor mounted below it.

The front axle was mounted on a subframe consisting of a pair of cast steel cross members, joined by gusset plates.

Towing attachments were provided, rigidly mounted at the front and rear.

Cab and bodywork

Where the FV1000 was unashamedly ugly, from some angles, the cab and body of the FV1200 was actually quite a stylish affair, with attractively-rounded, fully-enclosed front wheel arches, gently-curved engine compartment covers, and a functional three-door cab (two doors on the offside, one only on the nearside), intended for 12 men. The photographs of the two prototypes show a considerable number of detail differences between the two examples.

The twin-skin insulated cab was fitted with the normal two-piece windscreen design of the period, but without opening lights; one prototype appeared to have been fitted with the driver's windscreen wiper above the screen, and the passenger's, below. Tubular ladders provided access to the cab doors.

The rear wheels were covered by simple flat panelled splash guards with side valances; these mudguards were increased in width by the addition of infill panels when twin rear wheels were fitted.

The rear cargo body was of composite wood and steel construction, and was fitted with tubular steel hoops to

TUGS OF WAR

support a canvas cover. There was a hinged tailgate, provided with folding tubular metal access ladders for the convenience of the crew.

Care should be taken when examining certain period photographs of the first FV1201 prototype: the vehicle was delivered without a rear body and a number of photographs seem to exist with the body air-brushed onto the photograph. Admittedly it is well done, but it is not correct in every detail.

Electrical system
The electrical system was rated at 24V, and wired negative earth; all wiring and electrical devices were screened against radio interference. Four 12V 60Ah batteries were fitted, charged by an engine-driven CAV generator rated at 40A maximum output.

Winch
The prototype vehicle was fitted with a 15/20 ton capacity mechanical winch installed in such a way as to allow either front or rear pulls; long fairlead rollers were mounted at the extreme rear of the vehicle.

The rope diameter was 25mm, and the total length was 137m.

Radio equipment
The FVRDE specification called for provision to be made for installing 'Larkspur' C40 or B40 sets, 'or subsequent equivalent'.

DOCUMENTATION

Technical publications
FVRDE Specifications:
FV1201, artillery tractor, 30 ton, CT, heavy, 6x6.
FV1204, 30 ton tank transporter and semi-trailer.
FV1205, heavy 6x6 breakdown vehicle.

Prototype inspection record: FV1201(A), artillery tractor, 30 ton, CT, heavy. FVRDE, February 1955.

FV Range of vehicles. Information brochure: Part 2, 'B' vehicles. FVRDE, October 1951.

SCAMMELL
CHAPTER 5

CHAPTER 5: SCAMMELL

The Scammell company was originally founded in London's Spitalfields district as a firm of wheelwrights and coachbuilders, but the birth of the Scammell truck as we know it can be traced back to the 1921 Olympia Motor Show where G Scammell & Nephew exhibited an articulated six-wheel chassis with a 7.5 ton payload rating. Proving something of an innovation with what its manufacturer described as '3 ton operating costs', the truck quickly attracted more than 100 orders and led to the creation of a new firm, Scammell Lorries Limited.

By 1922, the level of sales was such that the company had outgrown its existing premises, and a move was made to Tolpits Lane, Watford in premises which Scammell were to occupy until their closure in 1987.

Scammell was not exclusively a constructor of heavy trucks, and although for example the company produced the distinctive three-wheeled 'mechanical horse' for a period of some 30 years, it is with 'heavies' that the Scammell name is most closely associated. The first of these was the 'Pioneer', introduced in 1927, and continuing in military service in tank transporter, and artillery and recovery tractor form throughout WW2 and well into the next three decades.

With its powerful, yet plodding Gardner 6LW diesel engine, massive Goodyear tyres and innovative front and rear suspension systems, the 6x4 'Pioneer' must have been something of a revelation in the late 'twenties. The walking beam gear cases at the rear, and pivoting front axle, allowed massive axle articulation, ensuring that all six wheels remained in contact with the ground regardless of how rugged the terrain. And that distinctive 'coffee pot' radiator was designed to ensure that no matter what the angle of operation, the tops of the radiator header tubes would not be uncovered. Even by today's jaded standards, the 'Pioneer' remains an impressive beast.

However, despite giving sterling service in all of the WW2 theatres of operation, by the end of the war it must have been clear that the days of the trusty 'Pioneer' were numbered. In the early 'fifties the Ministry of Supply began to take delivery of the 6x6 'Explorer'. Similar in approach, and retaining most of the features of the 'Pioneer', the 'Explorer' offered a higher top speed by virtue of the six-cylinder Scammell-Meadows petrol engine, and improved cross-country performance through the use of all-wheel drive. The 'Explorer' was only ever supplied as a heavy recovery and towing tractor, and was rated variously at 13 and 10 tons.

Shortly after deliveries of the 'Explorer' began, the Ministry of Supply placed orders with the company for a number of 20 ton 6x6 'Constructor' tractors, either with ballast bodies for towing full-trailers, or with a fifth-wheel coupling for use with the 30 ton Royal Engineers' semi-trailer. The 'Constructor' was equipped with conventional rear transmission and suspension using a bogie axle, and was powered by either a Scammell-Meadows petrol engine, or a Rolls-Royce C6 diesel engine.

In 1955, although no outward changes were made, Scammell was absorbed into the Leyland empire, and by this time, the last of the 'Pioneers' was definitely on its way out. Unfortunately for Scammell, who had been in financial trouble before the takeover, the British Army had standardised on the Thornycroft 'Mighty Antar' as a tank transporter, and further (British) military orders were not forthcoming.

Although Thornycroft itself had been taken over by Leyland in 1962, and ironically production had eventually been transferred to the Scammell works at Watford, it wasn't until the 'Antars' in turn were being pensioned-off that Scammell once again had the chance of supplying tank transporters.

By the 1970's, the 'Explorers' and 'Constructors' were getting a little long in the tooth, and certainly most of the latter had been disposed of within 15 years or so of their introduction. There were no military orders forthcoming for heavyweights, during the 'seventies and 'eighties. However, with the delivery of a number of medium-mobility 'Crusaders' and S26 vehicles bodied for the recovery and DROPS cargo roles respectively, the Scammell name continued to be represented in the British Army.

But the big prize was the 'Commander' 65 ton tank transporter intended to replace the 'Antars'. Prototypes appeared in 1978, with production versions being delivered in 1984 which means that the mighty Rolls-Royce engined 'Commander' is outside the scope of this book. However, despite the demise of Scammell as a separate name since the DAF takeover of Leyland in 1987, it's nice to know that Scammell trucks will remain in uniform for some time to come.

But... powerful and efficient though the 'Commander' may be, anyone with more than a passing interest in military vehicles will surely agree that there is a very special magic associated with the names 'Pioneer' and 'Explorer'.

CHAPTER 5.1

SCAMMELL PIONEER
R100, SV/1T, SV/1S, SV/2S, TRMU/20+TRCU/20
TRMU/30+TRCU/30

During WW1, motor transport was really still in its infancy and no purpose-made vehicles were available which could be considered suitable for transporting heavy tanks, and particularly for transporting them across country.

However, towards the end of the 'twenties, Scammell Lorries of Watford produced a prototype of what was to become the 'Pioneer', a powerful, six-wheeled heavy prime mover featuring an innovative suspension and transmission system which appeared to make the vehicle well-suited to off-road work. Available in both 6x4 and 6x6 configurations, the 'Pioneer' offered, for its time, a most impressive cross-country performance.

Scammell intended that the vehicle would be used in the emerging commercial heavy-haulage market, and rather like the Thornycroft 'Antar' which was yet to come (see Chapter 6.1), early examples were employed to haul large-diameter oil pipelines across the Jordanian desert. However, with its extraordinary power, walking-beam suspension, and massive axle articulation, the 'Pioneer' was obviously well-suited to the tank-transporter role, and Scammell themselves were keen to promote the vehicle for this application.

Unfortunately, the military authorities were slow to recognise that they had a need for such a vehicle.

For the duration of the Great War, and in the inter-war years, the Royal Army Service Corps had tended to rely on the railways for long-distance movement of tanks, assuming that the tanks would be able to travel under their own power from the railhead to the front line. Experience gained during WW1 had shown how vulnerable fixed railway lines could be, and the practice of making final delivery by road tended to result in an unacceptable rate of breakdown of the tanks themselves, which were not really suited to travelling on metalled road surfaces. Despite this, it was not until 1932 that the first 'Pioneer' was purchased for use as a tank-transporter tractor.

Equipped with a non-detachable fifth-wheel turntable for use with a purpose-designed 20 ton semi-trailer, the 'Pioneer' tank transporter was expected not only to be able to haul tanks across all kinds of terrain, but also to be able to recover tanks or other tracked vehicles which were unable to move under their own power. The loading arrangements were not ideal, and the trailer rear bogie had to be removed to enable the tank to be loaded, and

CHAPTER 5.1: SCAMMELL PIONEER

VEHICLE OUTLINES

R100

SV/2S

TRMU/20 and TRMU/30

TRCU/30

SCALE 1:100

SPECIFICATION

Dimensions and weight

	R100	SV/2S	TRMU/20	TRMU/30
Dimensions (mm):				
Length	6176	6176	6710	6710
with jib extended	-	7725	-	-
with semi-trailer*	-	-	16,950**	15,150**
Width	2593	2593	2615	2615
with semi-trailer*	-	-	2795	2882
Height to top of cab	2974	2974	2870**	3327**
Wheelbase:	4575	4575	4575	4575
rear bogies	1384	1384	1384	1384
20 ton semi-trailer	-	-	1397	1397
30 ton semi-trailer**	-	-	1550	1550
Track:				
front	2083	2083	2070	2083
rear	2159	2159	2020	2019
semi-trailer, outer	-	-	2172	2516
Ground clearance	324	324	324	324
Weight (kg):				
Laden	12,358	11,047	17,653	19,702
with semi-trailer**	-	-	36,043	50,705
Unladen	8528	9838	10,627	11,811
with semi-trailer**	-	-	15,782	20,262
Bridge classification***	16	16	16	16

Performance

Maximum speed, 38km/h.

Fuel consumption: 0.5-0.88 litre/km, on road. Maximum range, on road: 20 ton tank transporter, 352km; 30 ton tank transporter, 288km; artillery tractor, 568km; recovery tractor, 856km.

Turning circle: tractor only, 21.35m; TRMU/20 tractor and semi-trailer, 21.86m (left-hand), 22.265m (right-hand); TRMU/30 tractor and semi-trailer, 25.0m (left-hand), 24.71m (right-hand).

Maximum gradient: 22-33% according to variant.

Capacity

Maximum towed load, including semi-trailer: TRMU/20, 20,000kg; TRMU/30, 30,000kg.

* With ramps in the raised position.
** Figures refer to Scammell semi-trailer.
*** Classification for solo tractor.

subsequently refitted - all this, of course without the benefit of hydraulics.

The vehicle was considered something of a novelty, never really being used for its intended purpose, and remaining in service largely for training purposes. There was no series production until close to the outbreak of war, and it is said that in 1939, the British Army possessed only two of these tank transporter tractors.

Other branches of the services, however, had been taking things a little more seriously, and in 1937, Scammell had begun to supply the Royal Artillery with the 'Pioneer' R100 heavy artillery tractor, intended for towing and supplying the 7.2in howitzer.

At more-or-less the same time as the artillery tractor was produced, Scammell's design department came up with a modified, longer-wheelbase version of the tank-transporter tractor, intended for use with the original 20 ton semi-trailer. In 1939, a quantity of these was delivered, but the trailer was soon modified to allow more convenient loading over ramps, and the whole outfit was subsequently uprated to 30 tons.

Also in 1939, the same chassis was developed into the third of the 'Pioneer' heavy variants, the recovery tractor. The first versions, designated SV/1S and SV/1T, were fitted with a lifting jib which consisted of a collapsible steel-girder 'A' frame carrying a horizontal 'I' beam, but less than 50 were produced before it was replaced by the SV/2S which was equipped with a more-adaptable two-position 3 ton sliding jib crane supplied by Herbert Morris.

Despite its somewhat stately progress - top speed was never better than about 40km per hour - the trusty Gardner oil-engined 'Pioneer' was enormously successful and many were retained after the war on the strength of the TA, or were found other useful roles, remaining in service way beyond their normal 'sell-by date'. Though it was essentially a pre-war design, never for example acquiring front-wheel brakes, the 'Pioneer' continued to give valiant service for some years after the war was over.

By the early 'fifties, the 'Pioneer' tank transporter had effectively been replaced by the 'Antar', and by the middle of the decade, the writing was certainly on the wall for the artillery tractor. In 1957, the Major General Director of the Royal Artillery stated in a War Office memo that the '6x4 Scammell is unsuitable as a heavy gun tractor... (because it has) no front wheel drive... poor ground clearance... and towing block too low'. It was suggested that the AEC 'Matador' or Leyland 'Martian' tractors would be more-acceptable substitutes.

The SV2/S recovery tractor was the longest-surviving variant, and although it was intended to be replaced by the 'Explorer', many were used well into the post-war years. In fact, the 'Pioneer' SV2/S went on to become the longest-serving of any British military recovery vehicle, and a handful of examples actually made it through into the 'eighties.

DEVELOPMENT

Scammell had begun development of the 'Pioneer' in 1925, with P G Hugh working under the design direction of Oliver North. A prototype appeared the following year, and although the vehicle had yet to acquire the massive appearance of the classic military 'Pioneer', nevertheless the major features were already in place.

The prototype was fitted with enormous balloon tyres which, combined with the superb axle articulation, enabled the vehicle to cross most types of terrain without difficulty. The degree of axle movement was achieved by using a centrally-pivoted swing axle at the front, with a walking beam arrangement at the rear, allowing the rear wheels on either side of the vehicle to move relative to one another in the vertical plane by some 300mm.

The walking beams, which also served as gear cases for the final drive to the rear wheels, were pivoted on the chassis. A single differential transmitted the power to the pivot point of the gear cases. This arrangement, which incidentally was to remain almost unchanged on Scammell, and ultimately Leyland, heavy military vehicles for some 30 years, allowed the axles to attain extreme angles of articulation relative both to the chassis and to each other, without affecting the drive mechanism. Photographs exist which show early 'Pioneer' tractors travelling across rugged terrain with the wheels at seemingly-impossible angles. But of course, it was this feature which gave the 'Pioneer' its superb cross-country performance.

Although not featured on the prototypes, the 'Pioneer' soon acquired its other major recognition feature - that famous 'coffee pot' radiator. Forming part of the 'Still' tube cooling system, the 'coffee pot' was intended to provide a reservoir of water which would prevent the radiator tubes from becoming uncovered. The system was not pressurised and under most conditions wisps of steam could be seen drifting from the 'coffee pot'.

By 1928/29, Scammell were offering the 'Pioneer' in both 6x4 and 6x6 form, the latter literally providing wall-climbing performance. In 1932, Transport magazine published a report of a demonstration of a rigid six-wheeled 'Pioneer', of 6x4 configuration, which concluded '...mere description is inadequate to convey

CHAPTER 5.1: SCAMMELL PIONEER

a true impression of the performance of these remarkable machines'.

While commercial versions of the 'Pioneer' tractor employed Scammell's own petrol engine, all of the military versions were fitted with the 8.4 litre Gardner 6LW oil (diesel) engine, rated at 102bhp. Although obviously not a sparkling performer, this engine quickly acquired a reputation for reliability and performance, and was treated with considerable respect by users, particularly in the Western Desert.

Tank-transporter tractors

In 1932, the War Office purchased its first 'Pioneer' tank-transporter tractor, equipped with a 20 ton semi-trailer for use as a transporter for the Vickers medium tank. The design of the semi-trailer was similar to that of a modern low-loader, and the rear bogie had be removed to allow the tank to be loaded. Unfortunately, unlike a modern low-loader, this had to be done by means of manually-operated screw jacks and was inevitably a somewhat laborious process. Registered MV 5364 (WD number H22509), the vehicle remained in service until possibly the outbreak of war, but no further purchases were made for another seven years.

For some reason, the War Office had elected to purchase the four-wheel drive version rather than the six. Perhaps trials had shown that the 6x6 configuration offered greater complexity, but with little additional usable benefit. Although the 'Pioneer' was always offered by Scammell as a six-wheel drive version, the military authorities stuck resolutely with the four until the immediate post-war years when a 6x6 prototype version was produced. This was fitted with an early Rolls-Royce B80 engine, and used as a mobile test bed both for the engine, and for the 'Explorer' (see Chapter 5.3).

It was not until 1938 or 1939 that orders were placed for a reasonable number of fifth-wheel tractors intended for tank recovery and transporter duties. The first eight of the new tank-transporter tractors were rated at 20 tons and were supplied with the same type of semi-trailer as the original military 'Pioneer', with a non-detachable turntable coupling between the tractor and trailer. However, it was soon realised that if the trailer were to incorporate rear loading ramps which would allow the trackways to pass over the trailer wheels, loading would take considerably-less time.

The redesigned tractor and trailer, rated initially at 20 tons, but soon uprated to 30 tons with larger tyres and a new semi-trailer (designated TRMU/30 for the tractor, and TRMC/30 for the trailer), proved itself to be a most successful combination. Nearly 550 were delivered during

First Pioneer transporter, with fixed 20 ton semi-trailer, 1932 (TMB)

Same vehicle from rear showing tractor-to-trailer coupling (TMB)

Chassis for prototype R100 artillery tractor (TMB)

CHAPTER 5.1: SCAMMELL PIONEER

Experimental armoured car, showing use of rear tracks (Scammell)

Rear view of TRMU/20 with 20 ton semi-trailer (BCVM)

TRMU/30 with loaded 30 ton semi-trailer (TMB)

the years 1939-45, and many remained in service until the 'fifties.

Unfortunately, production facilities at Scammell's factory were restricted, and the artillery tractor was always given priority, which meant that there was a shortage of tank transporters through most of the war years. Production of the tank transporter never exceeded the peak rate of 17 vehicles a month achieved in 1943, and this continual shortage led to the use of other vehicles such as the Albion CX24 (see Chapter 2.1) and the Diamond T 980 (see Chapter 3.1).

Artillery tractors

Although considerable development work had gone into the tank transporter, the first military 'Pioneer' to be purchased in quantity was the R100 artillery tractor which began to be delivered in 1937. Despite having no rear mudflaps, and featuring the rather stylish eight-spoke wheels of the commercial 'Pioneer', the prototype (registered DMF 952) differed very little from the series production vehicles.

The R100 was a steel-bodied vehicle, with a well-type floor, designed for towing heavy artillery pieces; there was accommodation for nine men in the cab and rear body, together with a small rail-mounted loading crane for handling ammunition, which was also carried in the rear. An underfloor winch, rated at 8 tons, was provided for self-recovery.

Recovery tractors

Writing in his excellent book 'Breakdown', Brian Baxter of the REME Museum, reported that in 1928 a 'Pioneer' 6x4 chassis was purchased for trials with a purpose-designed mock-up armoured body. Once the trials were over, the body was apparently removed, and the vehicle converted to a breakdown truck; it remained at MWEE, Farnborough for many years.

However, it was not until close to the outbreak of war that a purpose-designed breakdown vehicle was purchased based on the 'Pioneer' chassis. Deliveries began somewhat hesitantly in 1939, but production was increased to a maximum of four vehicles per month, which was all that Scammell could manage at the time, following the evacuation of the BEF from the beaches at Dunkirk.

The first breakdown-recovery tractors were fitted with a collapsible steel-girder 'A' frame carrying a horizontal 'I' beam, a pulley was fitted at the far end for the recovery cable. Designated SV/1S and SV/1T, only 43 examples of this version were produced before it was replaced by the SV/2S, fitted with a Herbert Morris two-position sliding jib crane.

CHAPTER 5.1: SCAMMELL PIONEER

In all, some 1700 of these recovery tractors were produced, some of which were issued to the RAF and to other armies of the British Empire, but throughout the war the numbers available never matched the demand, and production was supplemented by American-built wreckers such as the Diamond T, Kenworth and Ward La France. 'Pioneer' breakdown vehicles remained in front-line service until well into the 'fifties, when they were finally superseded by the Scammell 'Explorers'.

Production
The total number of military 'Pioneers' produced was around 3000; some 550 examples were supplied as tank-transporter tractors, most were of the 30 ton type; 786 artillery tractors; and 1700 breakdown-recovery tractors.

It's not easy to say with certainty when the 'Pioneer' was finally superseded from production. Any outstanding military orders were most likely cancelled in 1945, and some commentators state that vehicles intended for the military were modified at the factory and eventually delivered to commercial operators - and of course, many surplus 'Pioneers' also found their way into civilian hands. Scammell were still showing a military-pattern tank transporter (described as 'Military 14'), and a ballast tractor variant (erroneously described as 'Explorer') in their 1954 catalogue, and as late as 1965, REME were still issuing EMER's for the breakdown-recovery tractor.

NOMENCLATURE

Tractor, 6x4, heavy artillery, Scammell; model R100
Steel-panelled artillery tractor designed to tow and supply the 7.2in howitzer. The vehicle could accommodate a crew of nine, with three in the cab, and the remainder seated in the interconnected rear body. The body also provided stowage for ammunition and vehicle stores, and a 10 cwt hoist was installed on a retractable overhead runway to assist with hitching the gun trail, and to help with loading ammunition. Fitted with an 8 ton mechanical winch, installed beneath the rear floor and equipped for rear pulls.

Tractor, 20 ton, GS, 6x4, Scammell; model R100/2
This was a post-war conversion of the TRMU/30 semi-trailer tractor, converted to a heavy recovery vehicle. The fifth-wheel equipment was removed and replaced by a conventional towing hitch. Other major modifications included the addition of six heavy-concrete ballast blocks on a steel frame positioned over the rear bogie, together with a small hand jib designed to help couple the trailer hitch.

Tractor, 6x4, breakdown, Scammell; models SV/1T, SV/1S
Early version of the heavy recovery tractor equipped with

Superb shot of TRMU/30 tractor and 30 ton semi-trailer (BCVM)

Close-up view of fifth-wheel coupling on 30 ton trailer (BCVM)

R100 artillery tractor (Scammell)

CHAPTER 5.1: SCAMMELL PIONEER

Same vehicle from rear - note winch rollers and cable (Scammell)

R100 towing WW2 6in howitzer (TMB)

Post-war, an R100 tractor awaits its fate (MP)

a hand-operated folding jib which could only be operated from one position; the jib was folded back into the body for travelling. An 8 ton mechanical winch was installed below the rear body, arranged for rear pulls only. The wooden-panelled rear body carried iron ballast weights, and was also used to house the various tools and pieces of equipment required for the heavy recovery role; ballast was also carried in a steel frame ahead of the radiator to help provide counter-balance for the front-end during suspended tows.

Only 43 of these vehicles were produced before the jib design was modified to the SV/2S version.

Tractor, 6x4, breakdown, Scammell; model SV/2S
Modified version of the SV/1T and SV/1S recovery tractor with a three-position Herbert Morris fixed jib, and equipped with a worm-driven hand hoist. The jib was rated at 2 tons at full extension, 3 tons in the midway position, and 8 tons fully retracted: later documentation only refers to the 2 ton and 3 ton positions so it may well be that the fully-retracted jib position was not really practical for use in recovery work.

Transporter, 20 ton, 6x4-8, semi-trailer recovery; model TRMU/20
Fifth-wheel tank transporter and recovery tractor intended for use with Scammell's own eight-wheeled 20 ton semi-trailer (TRMC/20), and equipped with a steel-panelled cab designed to accommodate seven men. Fitted with an 8 ton mechanical winch to help with loading disabled AFV's.

The first eight of these tractors were supplied for use with the original semi-trailer having removable rear wheels.

Transporter, 30 ton, 6x4-8, semi-trailer recovery; model TRMU/30
Development of the 20 ton tractor and trailer, fitted with larger-section tyres, and uprated to 30 ton capacity. During WW2, when Scammell TRMC/30 semi-trailers were in short supply, a number of these tractors were modified to allow them to also be used with the 30 ton Shelvoke & Drury semi-trailer.

Modifications and developments
Before the war, Scammell had tried to demonstrate the versatility of the 'Pioneer' chassis, even producing a 6x4 armoured car, and one vehicle with a radial engine installed over the rear axles.

During the war years, the emphasis was on production rather than development, and although the 'Pioneer' remained virtually unchanged during this time, inevitably, a number of minor modifications were made to suit specific roles.

CHAPTER 5.1: SCAMMELL PIONEER

An effective spade anchor was developed by the Experimental Beach Recovery Section of REME (EBRS), and fitted to a number of the recovery tractors; by reducing reliance on weight and tyre adhesion, this was intended to increase the effectiveness of the vehicle's prodigious pulling power. A few 'Pioneers' were fitted with armoured engine compartment covers and deployed on recovery work on the Normandy beaches.

In the early 1950's when prime movers were in short supply, some of the tank-transporter tractors were given concrete ballast bodies and a simple lifting jib, thus giving them a further lease of life (see R100/2, above). At least one artillery tractor was converted to a breakdown tractor, retaining elements of its metal rear bodywork, but receiving the jib and recovery equipment from an SV/2S recovery vehicle.

The 1970 edition of the RAOC tables of dimensions and weights, listed four variations of the SV/2S recovery tractor. The first was the standard recovery tractor, as described above. The second was a variation of this with attachments for towing the FV3751 mono-wheel fuel trailer normally towed behind the 'Centurion'. There was also a version equipped with the EBRS-designed spade-type earth anchor, and another equipped with both the trailer attachments and the spade anchor.

TRAILERS

The 'Pioneer' was designed for use with the following semi-trailers, and unlike modern fifth-wheel designs, the trailer was not readily uncoupled from the tractor:

TRMU/20 fifth-wheel tractor:
TRCU/20. Semi-trailer, 20 ton, recovery; manufactured by Scammell Lorries Ltd.

TRMU/30 fifth-wheel tractor:
TRCU/30. Semi-trailer, 30 ton, recovery; manufactured by Scammell Lorries Ltd.
Semi-trailer, 30 ton, recovery, 8TW; manufactured by Shelvoke & Drury Ltd.

Details of the Shelvoke & Drury trailer will be found in the chapter dealing with the Diamond T (Chapter 3.1).

Contemporary photographs also exist of the R100 artillery tractor towing the FV3621 20 ton low-loader trailer, and of the prototype for the post-war 6x6 'Pioneer', which was later to become the 'Explorer', towing what could be a prototype of the same trailer. Whether this was simply an expedient at FVRDE, or was a regular and normal occurrence is hard to say.

SV/1T recovery tractor with jib ready for use (BCVM)

SV/2S recovery tractor with two-position jib (TMB)

Same vehicle from rear showing jib in travelling position (TMB)

CHAPTER 5.1: SCAMMELL PIONEER

Pig-iron counterweights help hold down the front while lifting (NMM)

Artillery and recovery tractors had fairlead rollers at the front (NMM)

Artillery tractors converted to R100/2 ballast tractors using large amounts of concrete over the rear axle (REME)

DESCRIPTION

Engine

The commercial 'Pioneers' were fitted with Scammell's own petrol engine, but all of the military versions used the well-tried Gardner 6LW six-cylinder oil engine. Gross output was 102bhp at a governed 1700rpm, and the capacity was 8.4 litres.

The engine was equipped with a decompression device for manual starting, which entailed swinging the front draw bar to one side, and, to quote the Scammell manual, required three men, 'a man on the handle, and two on a rope... pulling the engine smartly over... it should then fire and commence to work'. I don't think I'd like to try it!

An ether carburettor was provided for starting at temperatures below -100°C. Small capsules of ether were fitted into a hand-operated puncturing device in the cab which introduced the ether into the engine air-intake system.

Fuel was supplied to the Gardner-designed injector pumps by an Amal diaphragm-type mechanical pump driven from the camshaft.

A small compressor was installed, driven by a power take-off on the left-hand side of the engine, in order to provide compressed air for the braking system.

Engine data
Capacity: 8396cc.
Bore and stroke: 4.25 x 6.0in.

Power output: 102bhp (gross); max engine speed 1750rpm; governed speed 1700rpm.
Maximum torque: 358 lbs/ft (gross).

Cooling system

The engine was water cooled, by means of what Scammell described as a 'Still' tube radiator, with the cooling water maintained at atmospheric pressure. The radiator consisted of offset rows of wire-wound tubes, presenting a massive cooling surface to the air; a shallow header tank was installed on top of the radiator... and sitting on top of the header tank was the famous 'coffee pot'... and on top of that was a simple water level indicator.

The idea of the 'coffee pot' was that regardless of the angle at which the vehicle tilted, the tops of the radiator tubes would never be uncovered.

Transmission

The drive was transmitted by means of a single 400mm diameter dry-plate clutch, in unit construction with the engine, to a Scammell six-speed and reverse, constant-mesh gearbox operated by dog clutches. An automatic brake was provided for the clutch to assist with smooth gearchanges.

The gearbox was mounted directly to the chassis using a three-point rubber suspension system, and connected to the clutch housing by means of a short shaft with muff couplings. In Scammell tradition, the gearchange pattern operated in an exposed, fixed gate. There were four neutral positions, with a dead point between most of the gears, and the handbook was at some pains to point out that a disabled vehicle should only be towed with the gearbox in the neutral position between third and fourth gears. The gear ratios were different on the tank-transporter version when compared to the artillery or recovery tractors.

The winch was also driven from the gearbox, and was indirectly controlled by means of the gearchange lever, which had to be put into the neutral position between second and third gears before the winch control lever could be used.

With a quoted top speed of just 38km/h, the gear ratios were obviously extremely low; overall ratio for first on the tank transporters, for example, was 181:1, and there was certainly no need for an auxiliary gearbox. Some idea of just how low-geared the 'Pioneer' was can be illustrated by the fact that the driver's handbook suggested the change from fifth to sixth gear, for example, should be made at around 10-14km/h, and it was stated that 'owing to the low ratio of first, second and third speeds, no recording is made on the speedometer'.

From the gearbox, an open Hardy Spicer propeller shaft ran the length of the vehicle to a single differential and drive-shaft housing unit which was used to drive both rear axles. Final drive to the rear wheels was by means of bevel-and-spur reduction gears to the walking-beam centres, with spur drive to the wheels through the walking beam gear cases.

The front axle was a steel tube, 90mm in diameter.

Axles and suspension

Suspension was provided for both front and rear axles, but no shock absorbers were fitted.

A single transverse elliptical spring was installed across the chassis at the front axle, pivoted centrally in a swivel housing, and pin-jointed to the offside stub axle swivel jaw; a slipper pad was used to support the spring on the nearside axle swivel. The axle was located laterally by means of an A frame. The total diagonal suspension movement from bump to rebound at the front was 610mm.

Suspension at the rear was by means of semi-elliptical springs, with one spring installed longitudinally under each chassis rail, arranged so that the gear cases were free to pivot whilst bearing on the spring ends. Total movement of the gear cases, measured at the wheel hub, was 300mm.

Semi-trailer
The semi-trailer bogie consisted of a pair of unsprung parallel balancer beams carrying one wheel at each end, pivoting on the chassis main channels to provide a walking beam configuration.

Steering gear

There was no power assistance on the steering which was provided by means of a Marles spiral cam-and-roller gear, with an overall ratio of 27.7:1 to give a turning circle of some 21m (22m on the 20 ton tank transporter, and almost 25m on the 30 ton version). The actual turning moment of the steering wheel was transmitted by means of a drag link to the offside wheel, and then by tie rod across to the nearside wheel.

A hand throttle control was mounted on the steering column.

Braking system

It seems extraordinary that on such a heavy vehicle there were no front brakes... but with a top speed of less than 40km/h, perhaps it was felt there was little chance of any need for a panic stop.

The air servo-assisted mechanical brakes operated on the rear wheels (ie, the driving wheels) only, by rods and compensating levers; air pressure was generated by an engine-driven compressor running at a maximum of 85kgf/m^2. If the air system were to fail, to quote a piece of classic understatement from the manual, the 'brakes on the motive unit can still be operated... but considerable force is necessary... and the brakes will not be very efficient'.

The main braking system was supplemented by a hand-operated 'Neate' brake, interconnected with the foot brake and operating on the rear wheels. This system, intended for use in an emergency, or as a supplementary braking system, applied the brakes by winding a flexible cable onto a drum by ratchet operation of the handle. The 'Pioneer' was not equipped with a separate hill-holder brake, and because the method of operation of the 'Neate' brake did not lend itself to a slow release, making hill starts with the transmission brake must have been a perilous affair.

A hand-operated transmission brake was also provided for parking, and for use in extreme emergencies.

Semi-trailer

The main braking system on the semi-trailer was operated by means of air pressure only, and was connected to the brake lines on the tractor. A separate mechanical parking brake was also fitted, with two separate mechanisms, each operating on one pair of wheels on the trailer bogie, by means of a hand wheel.

Road wheels

The wheels were 20in diameter, four-piece pressed-steel type, with enormous, normally cross-country type, tyres. The tyre size varied according to the application: on 30 ton tank-transporter vehicles, the tyres were 15.00x20 at the rear, with 13.50x20 on the front axle and on the semi-trailer; on the 20 ton tractors, the front tyres, and those on the semi-trailer were 10.50x20; while on the recovery and artillery tractor variants, the tyres were 13.50x20 all-round. The front (and semi-trailer) tyres were not always of the cross-country type.

Non-skid 'overall track' type chains could be fitted to the rear bogie wheels.

A spare wheel and tyre was carried in such a way as to permit (relatively) easy handling, and the brake compressor could also be used for tyre inflation.

Chassis

The chassis was of channel-and-box section design, consisting of two main channels running more-or-less straight from front to rear, with six cross members; the main channels were waisted at the rear bogie. The turntable, or fifth-wheel coupling, on the tank-transporters was bolted across the main chassis members over the driving wheels.

A substantial subframe of steel angle sections, was provided to carry the rear bodywork on the recovery and artillery tractors.

All versions were fitted with a spring-mounted draw bar at front and rear, each with a hook-type towing eye.

Semi-trailer

The semi-trailer chassis was constructed from steel channel sections, with cross members at regular intervals. The top surface of the main channels, which was arranged to slope upwards towards the fifth-wheel turntable, was formed to provide a trackway for the load. A wooden stowage bin was installed beneath the main channels.

Single- or two-part folding ramps were provided at the rear, the design being dependent on the trailer type and capacity.

Cab and bodywork

With its half-height doors and roll-up canvas side windows, the driver's cab of the 'Pioneer' was extremely basic.

Cab

All three vehicles shared the same sheet-steel three-man cab, engine compartment covers, and simple cycle mudguards; the rear bodywork varied according to the role. All versions included a large stowage bin beneath the cab on the offside, intended to hold the tracks for the rear bogie; on the nearside, this position was occupied by the fuel tank.

Front-hinged doors were provided for access from either side; some models seem to have been fitted with external door handles, others were not. The two-piece windscreen was designed to open, and glazed side windows were fitted to one side of the door; there were small glazed rear windows in the back of the cab on the recovery tractors. The cab was open at the rear on the tank-transporter and artillery tractor versions; on the tank transporter there was a canvas screen provided, while on the recovery tractor, there was direct access into the rear compartment.

The engine cover consisted of a fixed top panel, with hinged/removable side covers. The prototype artillery tractor was fitted with a stylishly-curved scuttle but this was replaced by flat panels once the cab was installed.

Because of the extraordinary axle articulation, the front mudguards were designed to move with the wheels, while the rear mudguards consisted of little more than flat steel mud flaps attached to the rear bodywork.

Bodywork

There was no rear bodywork as such on the tank-transporter version but the cab was extended sufficiently to accommodate four additional crew members.

The artillery tractor was fitted with a steel-panelled rear body designed to accommodate the ammunition and various items of stores as well as providing seating for another six crew members, while the rear body of the recovery tractor consisted of wooden-panelled lockers installed either side of the crane jib and intended to house the various items of recovery equipment.

Electrical equipment

The electrical system was rated at a nominal 12V, with a 24V starting circuit, and was wired negative earth. Four 6V 160Ah batteries were fitted, charged by a belt-driven CAV generator.

The starter, which was also supplied by CAV, was a type BS612/B48.

Winch

A 15 ton capacity Scammell-produced vertical-spindle mechanical winch was installed beneath the chassis, chain-driven by a power take-off from the auxiliary gearbox, and engaged and disengaged by means of a dog clutch. All of the winch controls were accessible from inside the cab. There was an engine cut-out device fitted to prevent overloading the winch.

Guide rollers and pulleys were provided to allow the winch rope to provide rear pulls at angles of up to 40° above and below the horizontal, and 90° from either side. The artillery and recovery tractor variants were also fitted with winch fairlead rollers at the front.

The winch, which was equipped with 137m of 25mm plough-steel wire rope, was intended both for loading equipment, as well as for assisting in travelling over bad ground, and thus provision was made for winching from the front or rear of the vehicle. The transmission arrangements allowed the vehicle to move under its own power whilst the winch was in operation.

A rigid 'A' frame device, described as a Hollebone yoke, was carried on the tank transporter vehicles to assist in loading operations. The yoke was stowed on the outside of the trailer locker.

DOCUMENTATION

Technical publications
As was the custom during WW2, most of the technical documentation was produced by Scammell themselves to their own specification. The familiar post-war EMER technical handbooks and repair manuals do not appear to exist for the 'Pioneer' range.

Driver's handbook
Transporter 30 ton, type 6x4-8 semi-trailer recovery. Scammell book nos 100/SL1A, 100/SL2.

Maintenance manual
Maintenance manual and instruction book. Scammell book no 101/SL1.

Instruction book
Tank transporter 30 ton, type 6x4-8 semi-trailer recovery, model TRMU/30/TRCU/30.

Waterproofing instructions
Waterproofing instructions for heavy artillery and breakdown tractors.

Parts lists
Tractor, 6x4, heavy artillery, model SV/2S.
Tractor, 6x4, breakdown, model SV/2S.
Tank transporter 30 ton, type 6x4-8 semi-trailer recovery, model TRMU/30/TRCU/30.

Meadows petrol engine type 6PC-630. WO code 17717.

Standards
Field inspection standard: recovery vehicle, medium, 6x4, Scammell. EMER R048.

Bibliography
Breakdown. A history of recovery vehicles in the British Army. Baxter, Brian S. London, HMSO, 1989. ISBN 0-11-290456-4.

British Army transport 1939-45. Part 1: tank transporters, recovery vehicles, machinery trucks. Conniford, M P. Hemel Hempstead, M.A.P Publications, 1972.

FV Range of vehicles. Information brochure: Part 2, 'B' vehicles. FVRDE, October 1951.

Military vehicle series. MV3: Scammell Pioneer. Conniford, Mike. Reading, Inkpen Art Productions, 1982. ISBN 0-907403-15-8.

The supertrucks of Scammell. Tuck, Bob. Croydon, Fitzjames Press, 1987. ISBN 0-948358-01-7.

CHAPTER 5.2

SCAMMELL CONSTRUCTOR FV12100 SERIES
FV12101, FV12102, FV12105

The post-war 'B Range' vehicle plan included a requirement for heavy, ultra-heavy, and super-heavy tractors in the weight ranges 10, 20, 30 and 60 tons. The 10, 30 and 60 ton categories were to be filled by the new CT purpose-designed vehicles, but the 20 ton category was assigned to the GS category, and was intended to be based on commercial designs. The only trouble was that there were no suitable commercial designs available; those that did exist, dated from before the war.

Writing in the minutes of a meeting held at the War Office on 21 July 1949 to discuss the production plan for the 'B' vehicles, the Director General of Fighting Vehicles (DGFV) stated that 'there was in existence a type of Scammell which would pull the 50 ton transporter... but only on a more or less level road'. Although he also went on to say that '...it would have nowhere near the performance of the post-war (FV1200) tractor', this was not the role for which such a vehicle was being considered.

It is hard to say for certain to which of the Scammell range he was referring, since at the time, the company was still offering the basically pre-war 'Pioneer' and had yet to launch any post-war models. Perhaps the War Office had got wind of the 'Constructor' which must have been in the development phase by then, for within a couple of years the Ministry of Supply had placed an order for a 'considerable number' of these vehicles in both ballast tractor and fifth-wheel configurations.

Despite the somewhat disparaging remarks made comparing the vehicle to the abortive FV1200, it was never required to pull a 50 ton trailer anyway, but it did have every appearance of being something of a panic purchase. The 'Pioneers' were getting on a bit, the FV1000 and FV1200 projects had been cancelled, the Thornycroft 'Antar' was yet to come fully to fruition, and there must have seemed a desperate need for a tractor for heavy engineering plant, and for the smaller armoured vehicles... and maybe the 'Constructor' just happened to be there.

The 'Constructor' had first appeared in 1952, and was an attempt by Scammell to replace the 'Pioneer', which by then was getting a little long in the tooth (although the two remained side-by-side in the Scammell catalogue until the mid-fifties). Originally equipped with the Meadows 10.53 litre diesel engine, the 'Constructor' was a purely-commercial design, subsequently modified for the military by the use of a Scammell-Meadows 6PC-630

TUGS OF WAR   117

CHAPTER 5.2: SCAMMELL CONSTRUCTOR FV12100 SERIES

petrol engine, and ultimately with the more-powerful Rolls-Royce C6NFL diesel.

Three versions of the 'Constructor' tractor were purchased by the Ministry of Supply.

The first vehicles to be delivered were the FV12101 Scammell-Meadows engined ballast tractors, designed for use with a 20 ton engineering plant trailer, and also incidentally, trialled for towing heavy artillery. This tractor was subsequently modified for use by the RAF, using a larger rear body and cab, and a Rolls-Royce C6NFL diesel engine, the redesigned tractor being designated FV12105. There was also the more-powerful FV12102 Rolls-Royce engined fifth-wheel tractor, intended for use with a rear-loading 30 ton semi-trailer, also primarily for engineering plant (FV3541).

Plans were also prepared for two artillery tractor variants (FV12103 and FV12104, the latter with a crane for barrel changing), and a ballast tractor for use with a 30 ton trailer. None of these variants progressed beyond the planning stage.

The 'Constructor' remained in service for some 20 or so years - both the FV12101 and FV12102 versions still appeared in the 1970 issue of the RAOC tables of vehicle weights and dimensions. Although it was a very capable machine, well-suited to off-road work because of the low ground pressure exerted by the dual rear wheels, it was confined to moving engineering plant and general stores. Scammell had demonstrated the 'Constructor' as a tank transporter tractor, and even sold some to overseas customers for this purpose, but this was not a role which it was destined to fulfil with the British Army.

DEVELOPMENT

By 1950, although it was still appearing in the Scammell catalogue, the trusty 'Pioneer' was really beginning to show its age, and the 'Constructor' was an attempt to remain true to the original concept but to bring it a little more up-to-date.

Introduced in time for the 1952 Commercial Motor Show, the six-wheel drive 'Constructor' lacked the powerful 'presence' of the earlier Scammells and, to some eyes at least, was a rather odd-looking vehicle.

The three-man cab, reputedly borrowed from an obsolete range of Bedfords, and certainly also used on the diminutive three-wheeled 'Mechanical Horse', looked like a hat that didn't quite fit. In fact, the cab was so small compared to the width of the chassis that on commercial versions, Scammell offered to offset the cab to the left or the right of the chassis according to the driving position. The simple 'cycle' type front mudguards were certainly

VEHICLE OUTLINES

FV11201

FV11202

SCALE 1:100

SPECIFICATION

Dimensions and weight

	FV12101	FV12102	FV12105
Dimensions (mm):			
Length	7579	7396	7660
Width	2517	2896	2896
Height	2972	3010	3280
Wheelbase	4804	4804	4804
rear bogies	1424	1424	1424
Track:			
front	2067	2110	2110
rear	1902	2029	2100
Ground clearance	346	409	409
Weight (kg):			
Gross laden weight	26,924*	61,065	26,926*
Unladen weight	14,427	15,038	17,022
Bridge classification:			
solo	33	24	36
with unloaded trailer	35	-	35
with loaded trailer	55	58	58

* Laden weight depends on ballast carried.

Performance
Maximum speed: on road, 59km/h; cross country solo, 16km/h; cross country laden, 10km/h.
Fuel consumption: FV12102, solo 1.04 litre/km, with trailer 1.89 litre/km; FV12102/FV12105, laden 0.58 litre/km.
Maximum range: FV12102, solo 608km, with trailer 338km; FV12102/FV12105, laden 805km.

Turning circle: left lock, 25m; right lock 24.5m.
Approach angle: 60°.
Departure angle: 60°.
Maximum gradient: FV12101, 25%; FV12102/FV12105, 33%.
Fording: unprepared, 762mm; prepared, 1983mm.

Capacity
Maximum towed load: FV12101, 31,565kg.

---

in the same tradition as those fitted to the 'Pioneer' and the later 'Explorer' but, with the height of the cab and bonnet, just didn't quite look right... and, anyway, for some reason there were at least three different patterns of mudguard fitted to the military versions.

However, with an adjustable driver's seat, wind-up windows and a fully-enclosed cab, the vehicle was arguably more modern in execution than the 'Pioneer', and there was sufficient power for the majority of commercial heavy-haulage jobs.

The original commercial 'Constructors' were fitted with a Meadows diesel engine, with a six-speed gearbox and two-speed auxiliary, or 'transposing', box. In combination with the 125bhp output of the standard engine, the resulting twelve speeds gave a credible performance. Unfortunately, the level of performance was not sufficient to impress FVRDE who in 1953, stated in their assessment report of the (commercial) prototype that the 'power to weight ratio is too low for Service requirements'. So, the first military 'Constructors' were fitted with a more-powerful 162bhp Scammell-Meadows petrol engine (FV12101 variants), and ultimately with the 175bhp Rolls-Royce C6NFL diesel.

Front suspension was very much in the pre-war Scammell tradition with a centrally-pivoted front axle suspended on a transverse spring. At the rear, there were longitudinal springs, with the axles arranged as a conventional tandem bogie. Drive to the rear axles was by conventional propeller shafts rather than by walking-beam gear cases.

The War Office had been considering various heavy vehicles since 1949, but with the difficulties which had arisen with the development of the two-purpose-designed CT tractors, the FV1000 and FV1200, there was some urgency to find suitable substitutes for the various roles. For a while, the situation must have become fairly desperate, and in 1953, a photograph appeared in Commercial Motor of a commercially-owned Scammell 'Highwayman' being trialled as a tank transporter on Canvey Island.

The 'Constructor' appears to have been one of a number of contenders, and was considered suitable for the lighter end of the heavy-haulage role - predominantly for trailers handling Royal Engineers' tracked plant, but also for moving smaller tracked AFV's, and possibly as an artillery tractor.

In 1952, the Ministry of Supply placed an order for what was described as 'a considerable quantity' of FV12101 20 ton 6x6 'Constructor' ballast tractors modified to meet FVRDE requirements. Unfortunately, it was not possible to make a 'service prototype' available for trials before production versions were delivered. Although this now

CHAPTER 5.2: SCAMMELL CONSTRUCTOR FV12100 SERIES

seems rather an odd thing to have done, perhaps it demonstrates the urgency of the situation which existed at that time.

Scammell made a commercial prototype vehicle available to FVRDE in order to prove the design as far as possible before deliveries began. The vehicle was given a protracted test, and in fact, it seems that there was more than one of these wooden-bodied prototypes since photographs exist showing at least two rear body designs, both with and without winch, with the vehicles running on commercial trade plates, registered 713 GC and 707 GC. With a couple of notable exceptions, it was stated that the vehicle was basically the same as the FV12101 tractor specified by FVRDE. The major difference was that the engine in the commercial prototype was a diesel version of the Scammell-Meadows unit, with a power output of 128bhp; the power output figure was increased to 162bhp on the military version (later to 180bhp), and of course, the military engine was a petrol unit. The other significant difference was the rear ballast body, which was of composite timber/steel construction.

Although the trials, carried out over some 16,000 kilometres, were generally satisfactory, the conclusions of the FVRDE report drew attention to four main areas. Firstly, there was felt to be a lack of power in the engine, but of course this had already been addressed by specifying a more-powerful version. The braking performance was felt to be inadequate and it was suggested that some attention was required in this area. The general reliability was felt to be satisfactory but with the caveat that 'extensive running of the Service prototype will... be necessary to assess its life mileage'. Lastly, some comment was made regarding the use of hydraulic steering dampers, which were not part of the original military specification, but which were to be carried across from the commercial versions.

The report concluded that 'the commercial prototype Scammell Constructor shews possibilities of meeting the Service requirements for vehicles serial nos FV12101 and FV12102, when fitted with the larger petrol engine and other features now asked for in the chassis specification'... which was just as well for deliveries were about to begin!

All of the vehicles were supplied under a single contract (6/Veh/7915/CB27a). The first vehicles to be delivered were the FV12101 ballast tractors which were covered by FVRDE Specification 9583, dated 10 November 1954. Deliveries began in 1954 at a price of £7350 each. Apparently the first few vehicles received were put through a user trials programme, and obviously the power level was still found to be wanting for before the FV12102 version was delivered later in 1954, the engine

Early Constructor commercial chassis under test (NMM)

'707 GC' - one of the commercial demonstrators tested by FVRDE (IWM)

... the other, registered '713 GC', seemed to have a longer wheelbase (IWM)

CHAPTER 5.2: SCAMMELL CONSTRUCTOR FV12100 SERIES

Production example of FV12101 ballast-bodied tractor on trials (BCVM)

An FV12101 crosses FVRDE's Centurion bridge (IWM)

The same vehicle was also tried with 'Rock Grip' tyres (IWM)

specification was changed and the Rolls-Royce C6NFL diesel was fitted. The FV12102 semi-trailer version had originally been described by the War Office as a 28 ton tractor, but this was not a standardised weight classification and before deliveries began, the rating was changed to 30 ton.

In 1958, the Rolls-Royce engine was also used in the RAF tractor, designated FV12105. This tractor was also fitted with a larger cab and body, and was not simply a re-engined version of the original FV12101.

All variants were fitted with a 15 ton winch.

According to contemporary records, the 'Constructor' was intended to be used to transport (tracked) engineering plant, bridge sections, and tracked AFV's etc by means of either a rear-loading 30 ton semi-trailer (FV3541), or a 20 ton full trailer (FV3621).

The '50 ton story' obviously persisted for some time. The catalogue for the 1966 exhibition of military vehicles held at FVRDE, Chertsey was still suggesting that the FV12105 variant was suitable for use with a 50 ton tank-transporter trailer... and although Scammell themselves claimed the vehicle was good for an all-up train weight of 100 tons, it must have been struggling a bit at that weight under anything other than ideal conditions.

Neither of the variations purchased was designated for use as an artillery tractor. However in a 1955 War Office memo discussing the possible future, or lack of, for the FV1200 tractor, it was stated that it had originally been felt that the 'Constructor' would be suitable for this role but the weight of the gun had risen to the point where the Director, Royal Artillery (DRA) believed that, at 18.5km/h, the average speed of the Scammell gun train was unacceptably low. It was also felt that the 'Constructor' lacked the power necessary to manoeuvre the gun into position under difficult ground conditions. The DRA had put up a case for producing the FV1200 for towing the 4in HAA gun, at an additional cost, compared to the Scammell, of £5 million. The Director, Weapons Development did not agree, stating that 'DRA should accept the inferior performance of the Scammell Constructor'.

There seems no other evidence that the 'Constructor' was ever actually used in this role and the planned purpose-designed FV12103 and FV12104 vehicles were not produced.

Production

A total of 346 units was supplied (274 FV12101, 52 FV12102, and 20 FV12105), and yet the vehicle remains something of an enigma. Perhaps it should simply be seen as a stop-gap between the WW2 tractors, such as the

CHAPTER 5.2: SCAMMELL CONSTRUCTOR FV12100 SERIES

'Pioneer', and the larger, more-modern vehicles which were to follow, like the 'Antar'. None seem to have survived into preservation, and very few photographs seem to exist of 'Constructors' in service.

Despite conducting trials of an export model 'Super Constructor' tank transporter with Dyson trailer in 1965, no other versions of the 'Constructor' served with the British Army and it was not until the advent of the 'Commander' tractor in the early 1980's that Scammell once again became suppliers of heavy vehicles to the British Army.

The FV12102 and FV12105 versions were still appearing in the 1966 Chertsey exhibition catalogue but had disappeared by 1972.

Although it's outside the scope of this book, there was also an interesting recovery variant produced for the New Zealand army, which was effectively a 'Constructor' cab and chassis with the rear body and recovery gear of the 'Explorer'... or perhaps it was an 'Explorer' recovery tractor chassis with a 'Constructor' cab.

On the commercial front, Scammell soon enlarged the cab of the standard vehicle, and went on to develop the 'Constructor' into a far more powerful and credible Mk 2 version, available with a crew cab and choice of Meadows and Scammell-Meadows petrol engines, and Gardner (yes, still the trusty old 6LW!), Rolls-Royce and Leyland diesel engines. In time, there were also the so-called 6x6 'Super Constructor' and 4x4 'Mountaineer' versions. None of these later versions were purchased by the Ministry of Supply.

NOMENCLATURE and VARIATIONS

FV12101. Tractor, 20 ton, GS, for full trailer, 6x6, Scammell Constructor
Original military 'Constructor' with permanent, steel ballast body intended for use with the FV3621 20 ton low-loader engineers' (RE) trailer. Fitted with Scammell-Meadows 6PC-630 10.53 litre petrol engine, six-speed main gearbox and two-speed auxiliary box. Simple three-man steel cab, generally with cycle wings at front; 12.00x20 tyres; steel rear body designed for loading with ballast, and also providing stowage for various items of equipment; 15 ton, Scammell mechanical winch installed beneath chassis.

FV12102. Tractor, 20 ton, GS, for semi-trailer, 6x6, Scammell Constructor
Fifth-wheel tractor designed for use with the FV3541 30 ton rear-loading engineers' (RE) semi-trailer. Fitted with Rolls-Royce C6NFL-140 four-stroke diesel engine, six-speed gearbox and two-speed auxiliary box. Simple three-man steel cab, seen with both cycle wings and

Production example of FV12102 fifth-wheel tractor (BCVM)

... perhaps the same vehicle from the rear - note the squared wings (BCVM)

Official factory portrait of FV12102 - again with squared wings (BCVM)

CHAPTER 5.2: SCAMMELL CONSTRUCTOR FV12100 SERIES

squared-off fully-enclosed front wings; 14.00x20 tyres; heavy-duty fifth-wheel coupling, with guide ramps, installed over rear axle; fitted with hydraulic equipment and couplings for operating the trailer ramps and jacks; 15 ton, Scammell mechanical winch installed beneath chassis.

FV12105. Tractor, 20 ton, GS, for 30 ton full trailer, 6x6, Scammell Constructor
Ballast tractor, similar chassis to FV12101, but with Rolls-Royce C6NFL-142A diesel; new, larger cab and cargo/ballast rear body design, usually with rounded fully-enclosed front wings; 14.00x20 tyres. Intended as a prime mover for 30 ton trailers, for use by the RAF; also stated to be suitable for use with the FV3601 50 ton full trailer.

Modifications
In about 1960/61, the Military Engineering Experimental Establishment (MEXE) at Christchurch in Hampshire, developed a 'Constructor' tractor for use as a high-speed road-surfacing machine.

The standard chassis was modified to accommodate a tank holding some 3000 litres of high-viscosity bituminous tar, and 12 tonnes of chippings. The tar was heated and delivered to the road surface from a spray bar, while the chippings were dispensed by gravity from a hopper. Although presumably the chippings would have been rolled to consolidate the surface, in this form the vehicle was able to lay road surfaces at speeds of 6.5-25km/h.

In 1974, the MoD also disposed of a 'Constructor'-based heavy recovery vehicle. This may have been a manufacturer's prototype, or may have been constructed by REME for a specific purpose. Possibly, it was the same vehicle that was photographed at FVRDE, heavily ballasted, and fitted with a demountable fixed-length jib designed to be erected on the rear of the chassis, together with a steel-framed winch housing immediately behind the cab.

TRAILERS

Available in ballast tractor and fifth-wheel configurations, the 'Constructor' was primarily intended for use with the following trailers/semi-trailers:

FV12101 ballast tractor:
FV3621. Trailer, 20 ton, 8TW/2LB, low loading; manufactured by Taskers of Andover (1932) Ltd, Hands (Letchworth) Ltd, and British Trailers & Co Ltd.

FV12102 fifth-wheel tractor:
FV3541. Semi-trailer, 30 ton, 8TW/2LB, rear loading, RE plant; manufactured by Taskers of Andover (1932) Ltd.

Excellent rear view showing fifth wheel and trailer connections (BCVM)

FV12102 with simple cycle wings - hitched to FV3451 trailer (IWM)

FV12105 chassis-cab under test at FVRDE (BCVM)

CHAPTER 5.2: SCAMMELL CONSTRUCTOR FV12100 SERIES

FV12105 ballast tractor:
FV3551. Trailer, 20 ton, 8TW/2LB, aircraft fuselage transporter; manufactured by Taskers of Andover (1932) Ltd.
FV3601. Trailer, 50 ton, tank transporter, No 1, Mk 3; manufactured by R A Dyson & Co Ltd (chassis), with Crane Fruehauf Trailers Ltd (body), (previously known as Cranes (Dereham) Ltd).
FV3621. Trailer, 20 ton, 8TW/2LB, low loading; manufactured by Taskers of Andover (1932) Ltd, Hands (Letchworth) Ltd, and British Trailers & Co Ltd.

DESCRIPTION

Engine
FV12102 vehicles
The early ballast tractors were fitted with a Scammell-Meadows 6PC-630 six-cylinder petrol engine, with a capacity of 10.35 litres. The engine was physically similar to the diesel version used in the original commercial 'Constructors', but with a power output of 162bhp (later increased to 180bhp). Engine speed was governed at a maximum of 2400rpm.

The engine was of the overhead valve type, with both inlet and exhaust valves installed in the cylinder head; the exhaust manifold was water-cooled to reduce temperature rise in the cab. Induction was by normal aspiration, with the engine breathing through a pair of large Solex 46WNHPO non-spill carburettors; petroleum Ki-gass starting equipment was also installed for those vehicles used in consistently-low temperature climates. Fuel was supplied to the carburettors by a David 'Korrect' mechanical diaphragm-type pump driven from the timing case.

At just 5.9:1 compression ratio, the engine was intended to run on 70 octane petrol, and ignition was by Scintilla NV6 vertical magneto, using KLG TMBR50 screened sparking plugs.

Lubrication was by conventional well-type sump, with a total oil capacity of 20.5 litres; an oil cooler was installed in front of the radiator.

A small compressor mounted on the left-hand side of the engine was used to provide compressed air for the braking system.

The commercial 'Constructor' was available with a direct-injection diesel version of this engine, designated Scammell-Meadows 6DC-630; although this engine had been used in the original trials vehicle, it was not installed in any of the military 'Constructors'.

FV12102/FV12105 vehicles
Despite the increase in power output to 175bhp,

Production example of FV12105 in RAF livery (KP)

Experimental Constructor-based tow/recovery vehicle (IWM)

Same vehicle with fixed jib stowed and carrying ballast weights (IWM)

experience with the FV12102 vehicle must have shown that the Scammell-Meadows petrol engine still lacked sufficient power for the intended purposes, and so the FV12102 and FV12105 vehicles were all fitted with a Rolls-Royce C6NFL-140 or C6NFL-142 direct-injection, four-stroke diesel.

Announced in October 1952, the C6NFL was a wet-linered in-line six of 12.17 litres, with a compression ratio of 16:1, giving a gross power output of 175bhp (184bhp for C6NFL-142), but with a near-20% increase in torque when compared to the petrol engine. The cylinder block casting was of modular design, enabling the flywheel housing to be mounted at either end, and the external accessories to be on either side; the code 'L' (or 'R') in the type designation was used to indicate which side of the block the injector pump was mounted. The valve configuration was overhead inlet and exhaust, with the valves driven by a single camshaft by means of conventional pushrods; there were two separate, removable cylinder heads.

Fuel-injection equipment was supplied by CAV, and a CAV mechanical governor held the engine speed down to 2100rpm. A separate Ki-gass (ether) carburettor was provided for extreme cold-weather starting.

Unlike the petrol engines, the C6NFL was a dry sump unit, with an oil capacity of 34 litres. Unusually, the engine featured both a hot-water oil heater system, and an air oil cooler. The oil heating system used engine coolant diverted from the radiator.

As with the petrol engines, a small water-cooled Clayton Dewandre compressor mounted on the left-hand side of the engine supplied compressed air for the braking system.

Engine data

	6PC-630	C6NFL-140/142
Fuel	70 octane petrol	DERV
Capacity (cc)	10,350	12,170
Bore and stroke (in)	5.125 x 5.125	5.125 x 6.0
Compression ratio	5.9:1	16:1
Firing order:*	153624	142635
Gross power output (bhp)	162-180	175/184
Engine speed (rpm)	2400	2100
Maximum torque (lbs/ft)	414-460	490

Cooling system
In both versions, the engine was water cooled, using a pressurised cooling system, with the relief valve operating at 2800kgf/m². The cooling water was circulated by means of a centrifugal belt-driven pump.

Like the 'Pioneer' and 'Explorer' tractors, the radiator was of the Scammell 'Still' tube type, but in this particular case without the 'coffee pot'; a belt-driven 610mm six- or eight-bladed cooling fan was installed behind the radiator.

Transmission
Drive was transmitted from the engine through a 400mm diameter (450mm diameter on Rolls-Royce engined vehicles) Borg and Beck single dry-plate clutch, to a rubber-mounted Scammell-designed gearbox, connected to the engine by a short propeller shaft. A combined transfer and auxiliary gearbox was mounted, integral with the main gearbox, providing final drive by multiple propeller shafts to the front and rear axles.

The main gearbox was of the constant-mesh type with dog engagement, giving six forward speeds and one reverse; gear changing was effected through the standard Scammell exposed gate. The gearbox had four neutral positions, with a dead point between most of the gears, and the handbook was at some pains to point out that a disabled vehicle should only be towed with the gearbox in the neutral position between third and fourth gears, not in the position between first and reverse since the gearbox lubrication system would not be operating correctly. A two-speed 'transposer', or auxiliary gearbox was provided, allowing the driver to select four or six-wheel drive in either auxiliary gear; the auxiliary ratios were 1.323:1, and 2.535:1.

All three models incorporated a winch power take-off on the transfer box, allowing winching at all main gearbox ratios; on FV12102 vehicles, the main gearbox also incorporated a power take-off for the semi-trailer jacking system.

The final drive system was extraordinarily complex, employing no less than five Hardy Spicer Type 1700 open propeller shafts; three transmitting the power from the transfer box to the rear axles, the other two driving the front axle.

Axles and suspension
Live axles were used at both front and rear, suspended on semi-elliptical springs; there were no shock absorbers. Aside from the steering components, the front and rear axles were interchangeable.

The front suspension consisted of a single, inverted multi-leaf transverse spring, centrally-pivoted on the main chassis cross member, with the spring ends secured to the axle by a pin joint at the right-hand end, and by resting on a slipper pad at the left-hand end. At the rear, longitudinal semi-elliptical springs were centrally-pivoted on brackets attached to a chassis cross member; the spring ends were attached to ball sockets bearing on the two bogie axles. The rear bogie allowed 425mm of diagonal articulation, or 305mm of vertical movement between the axles on each side.

Lateral location of the front axle was provided by a triangulated torque arm and ball socket attached to the chassis; at the rear, location was provided by Panhard rods The springs were not shackled to the axles at all, and were thus not subject to the torque reactions of braking or driving.

All three axles were of the double-reduction type, consisting of a spiral bevel pinion and wheel, and two sets of epicyclic gears. Drive to the front wheels was by Tracta-type constant-velocity joints; at the rear, the wheel hubs were fully-floating, with the power being transmitted through the hub caps, which were integral with the drive shafts.

Steering gear
A power-assisted steering system was fitted, consisting of Marles spiral cam-and-roller mechanism with air servo assistance provided from the braking system; hydraulic steering dampers were fitted to reduce kick-back and road shocks. Full steering control remained available in the event of failure of the servo system.

The wheel required 4.75 turns from one lock to the other, to give a turning circle of more than 25m.

Braking system
Combined mechanical/servo compressed-air brakes were provided on all three axles. A twin-cylinder compressor installed on the side of the engine and belt-driven from the crankshaft, was used to provide the compressed air, with an indicator gauge on the dashboard. A methanol anti-freeze system was provided to prevent moisture freezing in the braking system. The design of the braking system was such that the (mechanically-operated) rear brakes would still be available, albeit with some considerable effort on the driver's part, in the event of a compressor or servo failure.

Air line connections were installed at the front and rear of the vehicle to allow tandem operation with single-cab control, and to permit the trailer brakes to be connected to the tractor and operated via the vehicle braking system. A hand-operated reaction valve was also provided in the cab to permit independent operation of the trailer brakes, for example to prevent loss of control during steep descents.

A separate hand-operated 'hill-holder' brake was used to apply the brakes on all six wheels, intended for restarting on steep hills. A hand-operated 'Neates' brake operated on the rear four wheels only, by means of a multi-pull ratchet lever, providing both emergency braking, and a means of holding the vehicle at rest.

The cast-iron brake drums were 425mm diameter x 102mm width at the front, and 167mm at the rear.

Road wheels
The vehicle was fitted with single front wheels, with duals at the rear.

The wheels were manufactured by Dunlop, and were of the four-piece disc type. On FV12101 vehicles, the wheels were 8.37x20in, fitted with 12.00x20 cross-country tyres; on FV12102/FV12105 vehicles, larger wheels were fitted, 10.00x20in with 14.00x20 cross-country tyres (FV12105 also in service with 8.00x20in wheels mounting road tyres). Period photographs show vehicles equipped with 'Rock Grip' type tyres, as fitted to the 'Antar', run-flats (RF), and directional cross-country tyres.

Non-skid chains could be fitted to the front and rear wheels for additional traction.

A spare wheel was carried, either mounted on the left-hand chassis side on fifth-wheel tractors, or in the ballast box; a crane-and-davit system was provided for handling the spare wheel.

The brake system compressor could also be used for tyre inflation.

Chassis
The massive steel ladder chassis consisted of 305mm deep channel-shaped side members, waisted at the bogie, and boxed at the rear to provide additional stiffening, with six channel-section cross members. Hook type tow hitches were bolted to the front and rear cross members.

Fifth-wheel equipment
The fifth-wheel equipment, designed by FVRDE (FV252261) was of conventional pattern, with a manual lock for the trailer kingpin and twin ramps to guide the trailer onto the turntable. The equipment was mounted on a subframe bolted across the main chassis members.

Cab and bodywork
Cab
The three-man cab was of steel-framed, single-skin construction, with a short, tapered bonnet and typical imposing Scammell radiator guard; there were various shapes and patterns for the front wings. A double-skinned panel was fitted over the roof to help reduce heat gain in hot climates; this panel was extended forward over the windscreen to form a visor. The roof was also provided with a circular observation hatch and anti-aircraft hip-ring above the passenger's seat.

The right-hand side of the V-shaped, two-piece split windscreen was hinged at the top and could be opened fully; a small, sliding glazed window was provided in the cab rear bulkhead. Drop-down lights and swivelling quarter-lights were fitted in the lockable cab doors.

The cab floor consisted of removable timber panels, and there were two seats in the cab: a small, adjustable bucket seat for the driver, and a fixed bench seat for the two passengers.

The cab on the FV12105 vehicles was of a different pattern, looking something like that used on the contemporary Leyland 'Hippo'. The general design was more simple, with flat panels and a deeper two-piece windscreen.

Rear bodywork
On tractors intended for semi-trailers, a large, multi-compartment stowage bin was installed across the chassis, behind the cab. A large sheet-steel mudguard enclosed both rear wheels, terminating in a rubber mudflap.

The ballast bodies were based on the standard Scammell body but were modified by FVRDE Specification 9583. The body itself was constructed from 6.4mm thick steel, and was arranged in three compartments to house cast-iron weights and the various items of equipment which would have been carried on such a vehicle. A separate compartment was provided in the body to house the spare wheel, and a detachable davit was carried to enable the heavy wheel and tyre to be handled. The rear of the body was provided with an access door to the left-hand side stowage compartment, and a pair of side-hinged doors. Stowage compartments for the crew's personal kit were provided adjacent to the winch under the front compartment.

The rear body of the FV12105 vehicle was slightly longer, and was described as a cargo body, though it must certainly had facilities for carrying ballast. The rear doors extended to the full width of the body, with a tubular-steel ladder providing access into the body itself. Large sheet-steel and rubber mudflaps were fitted behind the rearmost axle.

Electrical system
The vehicles were wired on a 24V negative-earth system, using either four 6V 110Ah batteries on the petrol-engined vehicles, or four 6V 150Ah batteries on those vehicles equipped with a diesel engine. The batteries were housed in a lockable compartment beneath the cab.

A standardised FVRDE-designed 'No 2, Mk 1' or 'No 2, Mk 1/1' starter was installed on both types of engine.

On petrol-engined vehicles, the generator was an FVRDE 'No 1, Mk 1' or 'no 1, Mk 2/1' machine, producing 12A at 650rpm; the Rolls-Royce engines were fitted with a commercial CAV H5524/26CM generator with a maximum 20A output at 950rpm.

Winch
A 15 ton capacity Scammell vertical-spindle mechanical winch was installed beneath the chassis, chain-driven by a power take-off from the auxiliary gearbox. The winch, which accommodated 135m of 70mm wire rope, was engaged and disengaged by means of a dog clutch; an engine cut-out device was fitted to prevent overloading the winch. All of the winch controls were accessible from inside the cab.

Guide rollers and pulleys were provided to allow the winch rope to provide rear pulls at angles of up to 40° above and below the horizontal, and 90° from either side. The winch was intended both for loading equipment, as well as for assisting in travelling over bad ground, and thus provision was made for winching from the front or rear of the vehicle.

The winch drive was arranged to allow the vehicle to also move under its own power whilst winching.

DOCUMENTATION
Technical publications
Specifications
FVRDE Specification 9176. Chassis, tractor 20 ton, 6x6, GS, Scammell, FV12102.
FVRDE Specification 9220. Chassis and body, tractor 30 ton, 6x6, GS, Scammell, FV12105.
FVRDE Specification 9583. Body and mounting for tractor 20 ton, 6x6, GS, Scammell for full trailers, FV12101.
FVRDE Specification 9603. Body and mounting for tractor 20 ton, 6x6, GS, Scammell, semi-trailer, FV12102.
War Office Specification WO 70/Vehs/87. Tractor, 20 ton, FV12100 Series.

Report
FVRDE Report FT/B 295. Prototype report: Scammell Constructor 20 ton, 6x6, GS, (commercial version).

User handbooks
Provisional user handbook: tractor, GS, 20 ton, trailer, 6x6, Scammell.

User handbook:
Tractor, 20 ton, GS, 6x6, Scammell Constructor. WO code 17857.
Tractor, 30 ton, semi-trailer, GS, 6x6, Scammell Constructor. WO code 18371.

CHAPTER 5.2: SCAMMELL CONSTRUCTOR FV12100 SERIES

Servicing schedule
Tractor, 20 ton, GS, 6x6, Scammell Constructor. WO code 11086, 12523.

Parts list
Tractor, 20 ton, GS, 6x6, Scammell Constructor. WO code 19480, 17981.

Technical handbooks
Data summary:
Tractor, 20 ton, GS, 6x6, Scammell Constructor. EMER R170.
Tractor, 30 ton, GS, 6x6, Scammell Constructor. EMER R180.

Technical description:
Tractor, 20 ton, GS, 6x6, Scammell Constructor. EMER R172.
Tractor, 30 ton, GS, semi-trailer, 6x6, Scammell Constructor. EMER R182.

Engines, diesel, Rolls-Royce C range. EMER S532/1.

Repair manuals
Unit repairs:
Tractor, 20 ton, GS, 6x6, Scammell Constructor. EMER R173.

Engines, diesel, Rolls-Royce C range. EMER S533.

Field repairs: tractor, 20 ton, GS, 6x6, Scammell Constructor. EMER R174.

Base repairs: tractor, 20 ton, GS, 6x6, Scammell Constructor. EMER R174, Part 2.

Field and base repairs: Engines, diesel, Rolls-Royce C range. EMER S534.

Modification instructions
Tractor, 20 ton, GS, 6x6, Scammell Constructor; tractor, 30 ton, GS, semi-trailer, 6x6, Scammell Constructor. EMER R177, instructions 1-13.

Engines, diesel, Rolls-Royce C range. EMER S537, instructions 1-5.

Standards
Field inspection standard:
Tractor, 20 ton, GS, 6x6, Scammell Constructor. EMER R178, Part 1.
Tractor, 30 ton, GS, semi-trailer, 6x6, Scammell Constructor. EMER R188, Part 1.

Base inspection standard:
Tractor, 20 ton, GS, 6x6, Scammell Constructor. EMER R178, Part 2.
Tractor, 30 ton, GS, semi-trailer, 6x6, Scammell Constructor. EMER R188, Part 2.

Engines, diesel, Rolls-Royce C range. EMER S538, Part 2.

Miscellaneous instructions
Tractor, 20 ton, GS, 6x6, Scammell Constructor. EMER R179, instructions 1-6.
Tractor, 30 ton, GS, semi-trailer, 6x6, Scammell Constructor. EMER R189, instructions 1-2.

Engines, diesel, Rolls-Royce C range. EMER S539, instructions 1-9.

Tools and equipment
Table of tools and equipment for 'B' vehicles: tractor, 20 ton, GS, 6x6, Scammell Constructor. WO code 18337, table 1092.

Bibliography
Scammell builds a new 100 tonner. London, Commercial Motor magazine, 26 September 1952.

The supertrucks of Scammell. Tuck, Bob. Croydon, Fitzjames Press, 1987. ISBN 0-948358-01-7.

CHAPTER 5.3

SCAMMELL EXPLORER FV11300 SERIES
FV11301

By 1945, the Scammell 'Pioneer' was effectively more than 20 years old, and whilst these days this might mean that a commercial vehicle is nearing the end of its production cycle, it would certainly not suggest that it was close to being obsolete. But back then, the pace of change was more rapid, and 20 years represented a significant period in automotive development, even more so since the necessities of war had accelerated technological change by a considerable rate.

The 'Pioneer' had proved itself to be reliable, and was well-liked by the user arms, but it was beginning to show its age. Although all three of the variants remained in service during the immediate post-war years, the War Office was keen to replace them with something a little more modern.

The post-war 'B range' vehicle plan had included 10 ton breakdown-recovery tractors in both CT and GS categories, which were intended to supersede the SV/2S 'Pioneer'. The War Office was looking for something which would be able to offer the same virtues of tough, functional simplicity that had made the plodding Scammell such a success.

Enter the 'Explorer'... son of 'Pioneer', and the contender for the GS breakdown-recovery tractor role, with the CT role being filled by the Leyland 'Martian' (see Chapter 4.2).

For some time Scammell had offered both 6x4 and 6x6 versions of the 'Pioneer' chassis in its catalogue. The War Office had elected to buy only the 6x4; the 6x6, which was not actually popular with many customers, was ultimately withdrawn. However, towards the end of the war, FVRDE had conducted trials with a 6x6 version of the 'Pioneer' (incidentally fitted with a standardised Rolls-Royce 'B Series' engine in place of the Gardner diesel unit). These trials provided valuable experience in the behaviour of the Scammell chassis with all-wheel drive and were to lead directly to the development of the 'Explorer' recovery tractor, Scammell's first post-war military vehicle.

The 'Explorer' resembled the earlier 'Pioneer' in most respects but was of more compact proportions, with the front axle, now driven, set back under the engine, and with a squarer, taller radiator and engine compartment. A Scammell-Meadows 6PC-653 petrol engine was employed in both commercial and military versions, with many considering this a retrograde step, while the

CHAPTER 5.3: SCAMMELL EXPLORER

transmission and suspension arrangements were hardly changed from the trusty 'Pioneer'.

In the 1955 Scammell corporate brochure, the 'Explorer' was said to have been 'specially designed to provide exceptional axle articulation... which enables the machine to operate in conditions quite impossible for normal off the road wheeled vehicles'.

If this was true for the 'Explorer' it was also true for the earlier 'Pioneer' for there was no question of the vehicle having been completely redesigned. Both time and money were short, and Scammell were keen to secure post-war sales to both military and commercial customers... and anyway there was no advertising standards legislation in those days to check the truth of statements made in leaflets and advertisements.

The vehicle entered service in 1950, at a time when there was a desperate shortage of heavy recovery vehicles. The initial order was for 125, but the total in service eventually was to rise to more than 700.

Ultimately, the 'Explorer' was supplied in one major, and one minor variant, with only small differences between vehicles supplied under a total of six contracts. By far the most common was the recovery tractor, with its 15 ton winch and powered jib; less common, was the towing vehicle which had a similar appearance but was possibly not fitted with a jib. The latter tractors were mostly employed by the RAF, and there was also a civilian version, available in ballast tractor form.

There was no further development of the vehicle into variants for other roles, and the 'Explorer' remained in service, alongside the CT classified Leyland recovery tractor, until the late 'sixties and early 'seventies when the AEC 'Militant' Mk 3 based tractors began to come into service (see Chapter 1.2).

DEVELOPMENT

The 'Explorer' was a direct development of the 'Pioneer', with which it shared many features. Design work must have started in the late 'forties, and deliveries of the military versions started in 1950, with vehicles being issued to REME units during 1951. The stated role of the 'Explorer' was the recovery of wheeled vehicles and armoured cars up to the 10-ton class.

The chassis length remained more-or-less unchanged, but the axles were brought closer together to reduce the wheelbase by about a metre when compared to the 'Pioneer'. Unlike the 'Pioneer', which had the front axle placed well forward, the axle on the 'Explorer' was set back under the engine.

VEHICLE OUTLINES

FV11301

SCALE 1:100

The plodding, but reliable Gardner 6LW diesel was replaced by a more powerful, and of course thirstier, Scammell-Meadows petrol engine, and the vehicle acquired front-wheel brakes at last. The engine block and cylinder heads were based on the earlier compression-ignition model but the engine was given more torque and more brake horse power, albeit at the cost of a fuel consumption often no better than 0.1 litre/km. However, the petrol engine gave the vehicle a far more acceptable performance with acceleration to 48km/h taking just 30 seconds, and a useful top speed of more than 60km/h.

The combination of six-wheel drive, more power and better brakes gave the 'Explorer' enhanced performance both on and off the road. Commercial Motor magazine road-tested, or rather mud-tested, a commercial example in 1952 and, despite inadvertently carrying out some of

SPECIFICATION
Dimensions and weight

	Chassis numbers:		
	6998-7122	7255-7483 7844-7848	7849-8168
Dimensions (mm):			
Length	6260	6265	6291
Width	2635	2635	2635
Height	3162	3162	3162
Wheelbase	3507	3507	3507
rear bogies	1302	1302	1302
Track:			
front	2159	2159	2159
rear	2171	2171	2171
Ground clearance	343	343	343
Weight (kg):			
Laden	13,550	13,230	13,230
Unladen	12,190	11,870	11,870
Jib performance:			
Maximum lift (kg):			
inner position	3000	3000	3000
outer position	2000	2000	2000

Bridge classification: 14.
Performance
Maximum speed, on road: 63km/h.
Average speed, off road: laden 19km/h.
Fuel consumption: 0.63 litre/km, on road.
Maximum range, on road: 502km.
Turning circle: 18.3m.
Maximum gradient: 44%.
Approach angle, 60°; departure angle, 60°.
Fording: unprepared, 762mm; prepared, 1525mm.
Capacity
Maximum towed load: including semi-trailer, 15,000kg.

the test with the bogie brakes partly applied, were most impressed. The vehicle was apparently able to haul 32 tonnes up a 25% incline, with a successful stop-and-restart halfway up the incline. The magazine also reported that they were unable to find a gradient which could stop the 'Explorer' and stated that the vehicle was able to climb a 44% gradient in second gear.

The famous 'coffee pot' radiator remained but was made deeper, while the engine cover was much squarer in plan.

As with the 'Pioneer', the walking beam gear cases for the final drive to the rear wheels, were pivoted on the chassis, with a single differential transmitting the power to the pivot point of the gear cases. This transmission and suspension configuration allowed the axles to attain extreme angles of articulation relative both to the chassis and to each other, without affecting the drive mechanism, and without imposing undue stress on the chassis frame itself.

Although similar in appearance to the earlier 'Pioneer' cab, the 'Explorer' made at least some concessions to crew comfort, with its insulated double-skin construction, wind-up windows and full-height doors, and on some models hot-water heating and demisting. The rear body remained virtually unchanged, but the vehicle did incorporate a completely-new 4.5 ton, two-position power-operated jib running from a second mechanical winch.

The prototype vehicle, and vehicles with chassis numbers 6998-7122, and 7255-7381, did not feature the rearmost lockers which were normally installed either side of the spare wheel, and on certain vehicles, the retaining cage on the offside under-cab locker was replaced by chains.

Almost all of the production total was for the recovery vehicle variant, but contract 5924 called for nine vehicles (chassis numbers 7475-7483) for use by the RAF as a towing tractor, possibly without the jib fitted. Other, standard, tractors were also supplied to the RAF with chassis numbers 7457-7474.

There is no documentary evidence to suggest that a fifth-wheel tractor version was contemplated, although since the 'Pioneer' was constructed in this form, it would be logical that such a vehicle might have been considered. Photographs taken at FVRDE of FVPE project 4255 show a heavily-ballasted cab and chassis with the rear of the chassis cut away at an angle, perhaps to clear a semi-trailer chassis, and with what could be runways to the fifth-wheel turntable bolted across the main chassis members. The air receiver for the braking system was installed on top of the chassis, forward of what would be the turntable position, and the vehicle also appears to have no winch.

However, there were no variants produced in any quantity, and development during the vehicle's life amounted to just a few minor changes. There were alternative mountings for the front and rear draw bars; some vehicles show a spring-loaded draw bar, others have a solid mount. Photographs of vehicles at FVRDE show two distinct styles of cab roof, one with what appears to be an air deflector above the windscreen. Those vehicles equipped with heating and demisting equipment had cab roof air intakes; others were fitted with rear-mounted spade anchors; and some were provided with a rear tow hitches for the 'Centurion' single-wheel fuel trailer.

Vehicles with chassis numbers above 7122 were fitted with a 'hill-holder' brake in place of the transmission hand brake; the air pressure braking system was modified; there was no cab heating or demisting; and the engine

CHAPTER 5.3: SCAMMELL EXPLORER

was provided with magneto ignition rather than a coil and distributor system. These vehicles were not fully waterproofed for wading.

The RAF tractors, chassis numbers 7457-7474, were also fitted with additional vacuum brakes for use with vacuum-braked trailers.

Production

A total of 726 vehicles was supplied, to both the War Office and the RAF (27 vehicles only), under seven contracts; the largest contract called for 320 vehicles.

Total production for both the military recovery vehicle, and the commercial ballast tractor, was somewhere around 1500, although the commercial versions were not common in Britain.

The 'Explorer' remained in the Scammell catalogue until at least 1955 but by the end of its life was looking rather archaic itself, and was probably finding few commercial buyers. In military service, by the late 'sixties or early 'seventies, the vehicle was superseded by the AEC 'Militant' Mk 3 medium recovery tractor. By that time, the 'Explorer', which could trace its origins back to a 'twenties design, must have become something of an anachronism.

NOMENCLATURE

FV11301. Tractor, 10 ton, GS, recovery, heavy, 6x6, Scammell
Heavy recovery tractor equipped with a 4.5 ton jib, and a 15 ton mechanical winch, installed below the rear body. The wooden-panelled rear body was used to house the various tools and pieces of equipment required for the heavy recovery role.

FV11301. Tractor, 10 ton, GS, towing, heavy, 6x6, Scammell
Rare RAF variant (contract 5924, nine vehicles only), based on the standard recovery tractor but equipped with a rear ballast body for use as a heavy towing vehicle, possibly without the jib. Otherwise similar in appearance and specification to the recovery tractor.

The 1970 edition of the RAOC tables of weights and dimensions listed four variants of the 'Explorer'. Alongside the two described above, there was a version of the recovery tractor with fittings for towing the FV3751 mono-wheel fuel trailer normally used with the 'Centurion', and a version of the recovery tractor with a spade attachment.

TRAILERS

The 'Explorer' was not generally intended for use with a trailer, but might occasionally have been found towing the light recovery trailer loaded with a disabled vehicle or AFV:

The prototype 6x6 Explorer was obviously based on the Pioneer (IWM)

... but some models were fitted with a B80 Mk 2D engine (IWM)

'Sally' shows off her somewhat eccentric wing arrangements (BCVM)

CHAPTER 5.3: SCAMMELL EXPLORER

FV3221. Trailer, 10 ton, 4TW/2LB, recovery, light; manufactured variously by Crossley Motors Ltd, Rubery Owen & Co Ltd, and J Brockhouse & Co Ltd.

The RAF towing tractor might also have been used with the FV3551 20 ton aircraft fuselage recovery trailer.

DESCRIPTION

Engine

All 'Explorers' were fitted with a Scammell-Meadows 6PC-630 six-cylinder petrol engine, either Mk 3/3 or Mk 3/7, with a capacity of 10.35 litres. The change to petrol was effectively forced on Scammell by the War Office who were keen to standardise fuels across the 'B range' vehicle fleet, neither Scammell, nor many of the users, saw this as a step in the right direction.

The engine was of the overhead valve type, with both inlet and exhaust valves installed in the cylinder head. There were two separate heads, and 'dry', steel cylinder liners were press-fitted into the block to provide a durable thrust surface for the pistons.

Induction was by normal aspiration, with the engine breathing through a pair of large Solex 46 WNHPO non-spill vertical carburettors; petroleum Ki-gass starting equipment was also installed for those vehicles used in consistently-low temperature climates. Fuel was supplied to the carburettors by a David 'Korrect' mechanical diaphragm-type pump driven from the timing case. An interesting feature of this engine is that the exhaust manifold was watercooled to prevent undue heat gain in the cab.

With a low compression ratio of 5.9:1, the engine was intended to run on 70 octane petrol, and ignition was by Scintilla NV6 vertical magneto, or by coil and distributor (chassis numbers 6998-7122), using KLG RML60 or TMBR50 screened sparking plugs respectively. Engine speed was centrifugally-governed to a maximum of 2400rpm.

The distributor, dynamo and fuel pump were all gear driven from the flywheel end of the engine, and a small Clayton Dewandre compressor mounted on the left-hand side of the engine was used to provide compressed air for the braking system.

Lubrication was by conventional well-type sump, with a total oil capacity of 20.5 litres; a tube-type oil cooler was installed in front of the radiator.

Engine data
Capacity: 10,350cc.
Bore and stroke: 5.125 x 5.125in.
Compression ratio: 5.9:1.

The Explorer was also used for deep-water wading tests (BCVM)

Early Explorer in production form under test at FVRDE (IWM)

... the nearside stowage locker is open (IWM)

CHAPTER 5.3: SCAMMELL EXPLORER

Firing order: 153624.

Gross power output: 162-180bhp.
Engine speed: 2400rpm.
Maximum torque: 414-460 lbs/ft at 1400rpm.

Cooling system

The engine was water cooled, by the Scammell 'Still' tube radiator, with the cooling system pressurised to 2800kgf/m². The radiator consisted of offset rows of detachable wire-wound tubes, presenting a massive cooling surface to the air; a shallow header tank was installed on top of the radiator, with the so-called 'coffee pot' forming both the radiator cap, and a small reservoir for the header tank.

The 'coffee pot' was intended to prevent the tops of the radiator tubes from becoming uncovered, regardless of the angle at which the vehicle tilted.

Transmission

The drive was transmitted by means of a single 400mm diameter dry-plate Borg and Beck clutch, in unit construction with the engine, to a Scammell six-speed and reverse, constant-mesh gearbox, pressure lubricated by a pump, and operated by dog clutches.

The gearbox was mounted directly to the chassis using a three-point rubber suspension system, and connected to the clutch housing by means of a short shaft with a flexible rubber coupling. In Scammell tradition, the gearchange pattern operated in an exposed, fixed gate. There were four neutral positions, with a dead point between most of the gears, and the handbook was at some pains to point out that a disabled vehicle should only be towed with the gearbox in the neutral position between third and fourth gears; if this instruction were ignored, the gearbox would not be properly lubricated.

The main and recovery winches were driven by a power take-off on the main gearbox; an engine cut-out device was fitted to prevent overloading of either winch.

Unlike the 'Pioneer' which had no front-wheel drive, the 'Explorer' was fitted with a fixed ratio auxiliary gearbox, which provided straight-through drive to the rear axle, and selectable six-wheel drive at a reduction ratio of 1.43:1. With an overall first gear, in six-wheel drive ratio of 125:1, the gearing was nowhere near as low as the 'Pioneer', but then there was considerably more power available from the engine.

From the auxiliary gearbox, an open Hardy Spicer type 1700 propeller shaft ran the length of the vehicle to a single worm-and-wheel differential and drive-shaft housing unit which was used to drive both rear axles. Final drive to the rear wheels was by means of bevel-and-spur reduction gears to the walking-beam centres, with

FVRDE often filmed vehicles under test (IWM)

Civilian-registered test vehicle crosses FVRDE's Centurion bridge (IWM)

RAF-registered towing vehicle with no jib (BCVM)

CHAPTER 5.3: SCAMMELL EXPLORER

Same vehicle from rear showing steel ballast body (BCVM)

Was this a prototype fifth-wheel tractor? (IWM)

... look at the rear of the chassis and the position of the weights (IWM)

spur drive to the wheels through the walking beam gear cases.

A second propeller shaft ran forward from the auxiliary gearbox to drive the front axle.

Suspension and axles

Spring suspension was provided for both front and rear axles, but no shock absorbers were fitted. The design of the suspension was such that none of the stresses due to cornering, driving and braking were transmitted through the springs to the chassis.

A single transverse elliptical spring was installed across the chassis at the front axle, pivoted centrally in a swivel housing, and pin-jointed to the offside stub axle swivel jaw; a slipper pad was used to support the spring on the nearside axle swivel. The axle was located laterally by means of an 'A' frame. The total suspension movement from bump to rebound at the front was 610mm.

The front axle was of the double reduction type, employing a spiral-bevel crown wheel and pinion, and two sets of epicyclic gears; final drive to the wheels was via 'Tracta' constant-velocity joints.

Suspension at the rear was by means of semi-elliptical springs, with one spring installed longitudinally under each chassis rail, shackled to the axle by U bolts, and to the axle at one end by means of a pivot pin. The axles were free to pivot on the springs whilst bearing on slipper pads on the spring ends. Total movement of the gear cases, measured at the wheel hub, was 300mm.

The rear axle was of worm-and-wheel pattern, with a bevel-type differential. The axle was pressure lubricated by a rotor-type pump driven from the worm shaft; an oil-cooler was also provided, fed by water by-passed from the engine cooling system. Drive to the four rear wheels was by spur reduction gears housed in the walking beam cases.

Steering gear

Steering was provided by means of a Marles spiral cam-and-roller gear, with air-pressure servo assistance from a separate reservoir on the braking system compressed-air circuit. The overall steering ratio was 28.5:1 to give a turning circle of 18.3m; the steering wheel was close to 500mm in diameter.

A bevel-type differential box was installed immediately above the steering box, actuating Clayton Dewandre steering valves. The actual turning moment of the steering wheel was transmitted by means of a drag link to the offside wheel, and then by tie rod across to the nearside wheel.

CHAPTER 5.3: SCAMMELL EXPLORER

A hand throttle control was mounted on the steering column.

Braking system

The air servo-assisted brakes operated on all wheels, mechanically on the rear wheels, and directly on the front wheels; air pressure was generated by a belt-driven Clayton Dewandre compressor operating at 60,000kgf/m².

A 'two-line' system was fitted for trailer braking, with the trailer reservoir fed via a diverter valve; application of the tractor brake pedal also applied the trailer brakes. A hand-operated reaction valve was provided to allow the trailer brakes to be applied independently, for example to prevent jack-knifing during steep descents. Vehicles supplied to the RAF, with chassis numbers 7457-7474 were fitted with an additional vacuum-operated trailer braking system.

A timing valve was installed in the braking system to ensure that the trailer brakes were applied slightly in advance of the tractor brakes, thus preventing jack-knifing or breakaway.

Coupling heads were fitted at the front and rear to allow vehicles to be double-headed under single control; and on vehicles with chassis numbers 6998-7122 a change-over valve was fitted to allow trailer brake operation when reverse towing, or 'nosing'.

The main braking system was supplemented by a hand-operated transmission brake on vehicles with chassis numbers 6998-7122, and a hill-holder brake on all other vehicles. The hill-holder brake was interconnected with the foot brake, operating on all wheels.

The effective area of the steel brake drums was 400x75mm at the front, and 425x100mm at the rear.

Road wheels

The wheels were 20in diameter, divided-disc type, with 14.00x20 cross-country tyres, either with the distinctive run-flat (RF) tread pattern, or simple bar grips.

Non-skid 'overall track' type chains could be fitted to the rear wheels.

A spare wheel and tyre was carried on the extreme rear of the body in such a way as to permit (relatively) easy handling, and the brake compressor could also be used for tyre inflation.

Chassis

The chassis was of channel and box section design, consisting of two main channels running straight from front to rear. The channel sections were boxed in at the centre and rear for additional stiffening, and braced by

Early RN-registered Explorer (KP)

RAF-registered recovery tractor (KP)

Excellent rear view showing recovery equipment and winch (IWM)

four cross members, two of welded construction, and two bolted.

A substantial subframe of steel angle sections, was provided to carry the rear bodywork; the main winch and recovery equipment were attached directly to the chassis.

All versions were fitted with a draw bar at front and rear, each with a hook-type towing hitch; the draw bar was either of a solid pattern, or spring loaded.

Cab and bodywork
Although the 'Explorer' offered a better level of comfort than the earlier 'Pioneer', it was still a pretty basic vehicle with a very functional appearance, and with few concessions were made to aesthetics.

Cab
The sheet-steel cab, engine compartment covers, and simple 'cycle' mudguards were not dissimilar to the 'Pioneer', although the cab appeared to be taller and of reduced width. The major improvements made over the earlier model were the incorporation of 50mm thickness of 'Isoflex', or similar insulation in the, now twin-skinned panels used to construct the cab, and the inclusion of drop-down windows in the doors; the twin-skin configuration was not continued into vehicles produced under the second and subsequent contracts. The floor was of timber construction, with removable inspection panels.

A large stowage compartment was provided beneath the cab on the offside, intended to hold the tracks for the rear bogie; on the nearside, this position was occupied by the fuel tank.

Front-hinged, lockable doors, with drop-down windows, were provided for access from either side, and the cab was designed to seat a crew of three; the driver's seat was adjustable for height and reach, while a small bench seat was provided for the other two crew members. The cab was fitted with a two-piece windscreen, designed to open, and there were glazed side windows fitted behind the doors on either side; small glazed, sliding rear windows were installed in the back of the cab, protected by steel bars. Vehicles supplied under the first contract were fitted with fresh-air heating and demisting equipment, with the intake on the cab roof; possibly it is these vehicles which have what appears to be a small air deflector above the windscreen. A metal-covered observation hatch for an anti-aircraft gun was provided in the cab roof above the passenger seat.

The engine cover consisted of a fixed top panel, with hinged/removable side covers. An FVRDE-designed oil-bath air cleaner was mounted on the scuttle, either side of, and above the bonnet top.

Because of the extraordinary axle articulation, the front mudguards were attached to the brake backplates, and were designed to move with the wheels. The rear mudguards consisted of little more than flat steel mud flaps attached to the rear bodywork behind the rearmost axle.

Bodywork
The rear bodywork consisted of a series of composite steel-and-timber lockers installed either side of the crane jib, and intended to house the various items of recovery equipment. On some versions, separate stowage boxes were installed either side of the spare wheel on the back panel.

A fold-down tubular steel ladder provided access through the left-hand side of the body.

Recovery equipment
The jib could be extended by means of hand-operated winding gear into one of two positions; the inner position permitted a lift of 3 tons, the outer position 2 tons. Twin arms were used to support the rear end of the channel-type jib runway. The jib winch, supplied by Turner, was power-driven by means of a propeller shaft connected to the main vertical winch, and was fitted with a torque-limiting device to prevent overloading. The jib was rigged for two-fall reeving.

Electrical equipment
The electrical system was rated at 24V, and wired negative earth.

All wiring and electrical devices were screened against radio interference. Four 6V 110Ah batteries were fitted, charged by an engine-driven CAV generator rated at 20A maximum output at 950rpm.

Winch
A 15 ton capacity (10 ton capacity on chassis numbers 6998-7122, 7255-7456) Scammell vertical-spindle mechanical winch was installed beneath the chassis, chain-driven by a power take-off from the auxiliary gearbox, and engaged and disengaged by means of a dog clutch; an engine cut-out device was fitted to prevent overloading the winch. All of the winch controls were accessible from inside the cab.

The winch cable was 70mm circumference plough steel wire rope, total length 137m. Guide rollers and pulleys were provided to allow rear pulls at angles of up to 40° above and below the horizontal, and 90° from either side. The winch was intended both for loading equipment, as well as for assisting in travelling over bad ground, and thus provision was made for winching from the front or rear of the vehicle.

CHAPTER 5.3: SCAMMELL EXPLORER

The transmission arrangements allowed the vehicle to move under its own power whilst the winch was in operation.

DOCUMENTATION

Technical publications

User handbook
Chassis, 16 ton, GS, recovery, heavy, 6x6, Scammell; and chassis, GS, towing, heavy, 6x6, Scammell. WO code 17820.

Servicing schedule
Chassis, 16 ton, GS, recovery, heavy, 6x6, Scammell; and chassis, GS, towing, heavy, 6x6, Scammell. WO codes 10557, 10626, 13453.

Parts lists
Chassis, 16 ton, GS, recovery, heavy, 6x6, Scammell; and chassis, GS, towing, heavy, 6x6, Scammell. Contract 3724, WO code 17464; contracts 5067 and 5820, WO code 17562; contracts 5820 and 7443, WO code 18269.

Meadows petrol engine type 6PC-630. WO code 17717.

Technical handbooks
Data summary: chassis, 16 ton, GS, recovery, heavy, 6x6, Scammell; and chassis, GS, towing, heavy, 6x6, Scammell. EMER R150.

Technical description: chassis, 16 ton, GS, recovery, heavy, 6x6, Scammell; and chassis, GS, towing, heavy, 6x6, Scammell. EMER R152.

Waterproofing instructions
Recovery vehicle, medium, 6x6, Scammell. EMER R155, instructions 1-2.

Modification instructions
Tractor, GS, recovery, medium 6x6, Scammell. EMER R157, instructions 1-27.

Complete equipment schedule
Recovery vehicle, wheeled, medium, 6x6, Scammell (FV11301); recovery vehicle, wheeled, medium, 6x6, with spade, Scammell; recovery vehicle, wheeled, medium, 6x6, with mono-trailer attachment, Scammell. Army code 33836.

Tools and equipment
Tables of tools and equipment for 'B' vehicles: chassis, 16 ton, GS, recovery, heavy, 6x6, Scammell; and chassis, GS, towing, heavy, 6x6, Scammell. WO code 17897, Table 1003.

Bibliography
Looking at the Scammell Explorer. Freathy, Les. Axbridge, Somerset, Vintage Commercial Vehicle Magazine, May/June 1987.

Mud... water... dust... craters. Test of Scammell 6x6 Explorer. London, Commercial Motor magazine, 13 June 1952.

Breakdown. A history of recovery vehicles in the British Army. Baxter, Brian S. London, HMSO, 1989. ISBN 0-11-290456-4.

The supertrucks of Scammell. Tuck, Bob. Croydon, Fitzjames Press, 1987. ISBN 0-948358-01-7.

THORNYCROFT
CHAPTER 6

CHAPTER 6: THORNYCROFT

Founded by John Isaac Thornycroft in Chiswick as early as 1862, and established as a manufacturer of steam vans in 1895, the Thornycroft company was one of Britain's first commercial vehicle builders. In 1898, the company was set up at the Basingstoke premises where it was to remain for the rest of its working life, extending the production facilities many times during the next 70 years.

The company enjoyed a long association with the military, and a number of the early steam wagons were shipped to South Africa during the Boer war. Some 5000 of the company's 3 ton 'Type J' petrol-engined truck were purchased by the War Office during the Great War, and 20 or so years later, a similar number of 3 ton 'Nubians' served in Europe during WW2.

Between the wars, as well as continuing production of the successful cargo trucks, the company became well-known as a constructor of motor boats and motor bus chassis. This continued until 1948, when the ship-building and vehicle sides of the business were separated, and the latter was renamed Transport Equipment (Thornycroft) Limited.

The re-structured company was to remain at the Basingstoke site almost until its closure in 1969, following takeover firstly by AEC, or more correctly ACV, in 1961, and then the subsequent absorption of that organisation into the growing Leyland empire in 1962. Leyland had already owned the specialist Scammell company since 1955, and consequently there was something of an overlap with Thornycroft. Two specialist truck manufacturers were considered one too many and, towards the end of Thornycroft's life, some production was transferred to the Scammell factory in an attempt to rationalise the operations, and a number of Thornycroft trucks were built at the Watford site.

During the post-war years, the company had tried to carve itself a niche market in the supply of specialist commercial vehicles, concentrating on medium to medium/heavy chassis, and aiming them at fleet operators whose needs were slightly out of the ordinary. At times it seemed that there was a bewildering model range, with names such as 'Big Ben', 'Nippy', 'Nippy Star', 'Sturdy', 'Sturdy Star' 'Trusty', and 'Trident'. Operators were offered a choice of Thornycroft's own diesel or petrol engines, or the Rolls-Royce B80/B81 petrol units, and all-wheel drive was offered as an option on the 'Nubian' 4x4 and 6x6 chassis.

However, there are probably just two models for which the company is known amongst the ranks of military vehicle enthusiasts. The first is the ubiquitous 'Nubian', which in its 3 ton 4x4 and 5 ton 6x6 form served faithfully with both the Army and the RAF over a period of more than 25 years. The 'Nubian' can hardly be considered a heavyweight.

The second, is the 'Mighty Antar', a truly-massive 30 ton tractor which could be considered the spiritual successor to the ill-fated FV1200 30 ton Leyland tractor.

First launched on an unsuspecting public at the 1950 Commercial Motor Show, the 'Antar' was originally developed for transporting oil pipes in the Iraqi desert during the commercial exploitation of that country's oil fields.

The 'Antar' was the company's first foray into the really heavy end of the market, but it soon showed itself to be a massive success, attracting the attention of the British Army, and going on to replace the ageing Scammell 'Pioneer' and Diamond T tank transporters, both of which had been giving sterling service since WW2. With its eight-cylinder Rover 'Meteorite' petrol engine, or Rolls-Royce C8SFL diesel, the 'Antar' was assigned the unenviable task of hauling the Army's 65 ton 'Conqueror' tank from one training exercise to the next. By 1963, some 750 vehicles had been constructed, and the 'Antar' was in service with nine armies across the world.

True to its name, and judged by any standards, the 'Mighty Antar' was a most impressive machine. Standing 3100mm high, and with an overall width of more than 3000mm, most of which seemed to be occupied by radiators, the 'Antar' had enormous presence. In its original ballast tractor form, the vehicle was equipped with a 285bhp 'Meteorite' petrol engine, and was intended for use with trailers with a GVW up to about 68 tons. Fifth-wheel versions were also produced, subsequently with a 333bhp Rolls-Royce diesel, designed for use with semi-trailers with a GVW up to 66 tons. Everything about the 'Antar' was big... and slow. Flat out, the 'Antar' could scarcely manage 50km/h and it took getting on for a minute-and-a-half to get there!

The 'Antar' remained in service with the British Army until the mid 1980's when it was gradually replaced by the Scammell 'Commander'. By this time, the old-established firm of Thornycroft had been dead and buried for close to 20 years.

CHAPTER 6.1

THORNYCROFT ANTAR FV12000 SERIES
FV120001-FV12007

Despite having been in business since 1895, for the first 50 or so years of its existence, the heaviest vehicle that the Thornycroft company had produced was rated at a modest 12 tons, and most of its products were in the 3 and 5 ton categories. Then, almost out of the blue came the 'Mighty Antar', an ultra-heavy prime mover which had been purpose-designed for hauling steel oil pipelines across the deserts of Iraq. With its Rover 'Meteorite' engine and all-round massive construction, the 'Antar' was quite capable of hauling loads up to 85 tons weight under the most difficult working conditions.

Initially a commercial venture, the first 'Antars' were constructed to a specification laid down by contractors Geo Wimpey, and were delivered to the Iraq Petroleum Company. The vehicle proving trials had been carried out at the FVRDE test track at Chertsey, which of course was close to Thornycroft's Basingstoke works, more-or-less at the same time that FVRDE were carrying out the design studies for the abortive FV1000 and FV1200 projects (see Chapters 4.1 and 4.3). Obviously the military authorities started to take an interest in the 'Antar' as a possible tank transporter. After all, here was a proven design where most of the initial design work had already been carried out, and more to the point, carried out at Thornycroft's expense.

In use, these commercial 'Antars' acquitted themselves extremely well in the hostile Iraqi and Syrian deserts, with the vehicles eventually covering a collective total of more than 1,000,000 kilometres without a total breakdown. This compared most favourably with the trials of the purpose-built CT vehicles which had also been going on at Chertsey, and FVRDE was forced to pay very close attention. It was not long before the first military orders were placed.

The first military 'Antar' was the FV12001 Mk 1 steel-bodied ballast tractor which entered service in 1951. The tractor was intended for use with the FV3601 full trailer, produced by either Cranes or Dyson (see Chapter 7), and although the outfit was nominally rated at 50 tons, it was more than capable of hauling the 65 ton 'Conqueror' for which the massive FV1000 tractor had originally been planned.

The Mk 1 was quickly joined by the Mk 2, available in full-trailer and semi-trailer variants (FV12002/FV12003), which was uprated to a nominal 60 tons. The Mk 2 versions were trialled with both the FV3001 semi-trailer, and the redesigned FV3005 version during 1953-55.

TUGS OF WAR

141

CHAPTER 6.1: THORNYCROFT ANTAR FV12000 SERIES

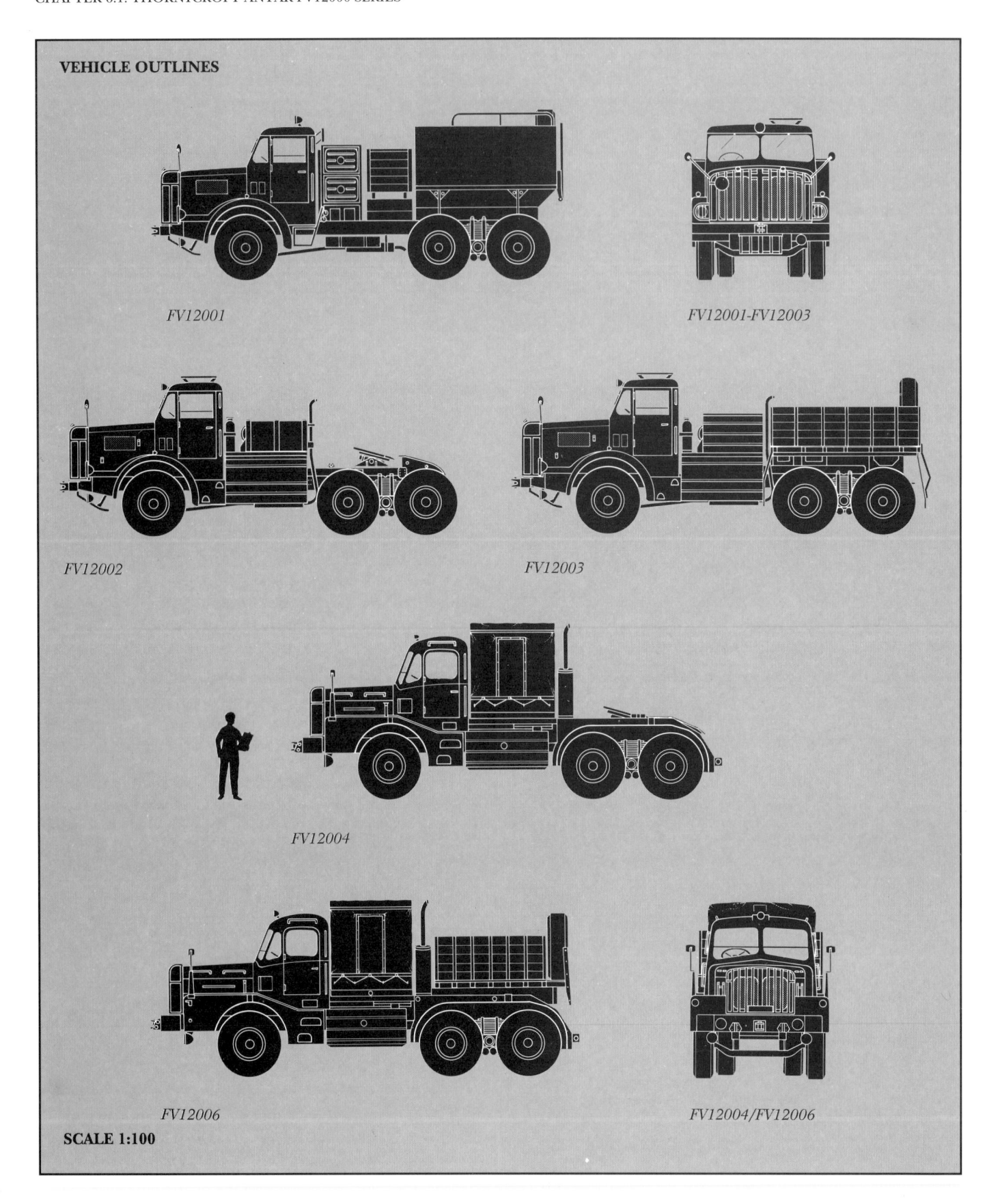

VEHICLE OUTLINES

FV12001

FV12001-FV12003

FV12002

FV12003

FV12004

FV12006

FV12004/FV12006

SCALE 1:100

CHAPTER 6.1: THORNYCROFT ANTAR FV12000 SERIES

SPECIFICATION

Dimensions and weight

	FV12001	FV12002/3	FV12004/6
Dimensions (mm):			
Length:			
ballast tractor	8440	8440	8700
fifth-wheel tractor	-	8150	8700
Width	2820	2820	3200
Height	3050	3050	3200
Wheelbase	4720	4720	4880
rear bogies	1580	1580	1580
Track:			
front	2260	2260	2250
rear	2290	2290	2290
Ground clearance	390	390	390
Weight (kg):			
Ballast tractor:			
gross laden weight	34,500*	35,590	35,590
unladen weight	20,100	20,100	23,440
Fifth-wheel tractor:			
gross laden weight	-	51,800	52,487
unladen weight	-	19,630	21,919
Bridge classification	35	52	53

* Laden weight depends on ballast carried; quoted figure includes 15,000kg of ballast (maximum 27,500kg).

Performance
Maximum speed: Mk 1 and 2 vehicles, 45km/h; Mk 3/3A vehicles, 52km/h.

Fuel consumption: Mk 1 and 2 vehicles, 2.25 litre/km; Mk 3/3A vehicles, 1.38 litre/km.

Maximum range: Mk 1 and 2 vehicles, 400km; Mk 3 vehicles, 702km.

Turning circle: 19.22m.
Maximum gradient: 18%.
Approach angle: 30°.
Departure angle: Mk 1 ballast-bodied vehicles, 29°; Mk 2 ballast-bodied vehicles, 50°; Mk 1 and Mk 2 fifth-wheel tractors, 90°.

Fording: unprepared, 760mm.

Capacity
Gross train weight: Mk 1 and 2 ballast-bodied vehicles, 142,550kg; Mk 2 ballast-bodied vehicles, 142,550kg; Mk 1 and Mk 2 fifth-wheel tractors, 101,820kg; Mk 3 ballast-bodied vehicles, 112,000kg; Mk 3 fifth-wheel tractors, 112,000kg.

These figures are based on published War Office material, and the apparent disparity between the gross train weights of the Mk 1 and 2 vehicles at 142,550kg, and the Mk 3 at some 40,000kg less must be attributed to a growing conservatism on the part of the authorities as the years passed.

The final version was the Rolls-Royce C8SFL supercharged diesel-engined Mk 3/3A, which first appeared in 1958, with deliveries beginning in the early 1960's. With a redesigned cab and frontal appearance, and new engine and transmission, the Mk 3/3A was again available in versions for both full and semi-trailers (FV12004/FV12006).

It was probably the proven performance, relatively low price (a commercial 'Antar' tractor was priced at £9000-10,000), and appetite for hard work and abuse of the 'Antar' which put the final nail in the coffin of the FV1000/FV1200 Series. Nearly 200 'Antars' were delivered to the military over a more than 10-year period, gradually replacing the Diamond T 980/981 and Scammell 'Pioneer' tractors/prime movers, both of which dated back to WW2.

Rather like the Alvis 'Stalwart' which was also a successful private venture, the 'Antar' - it was never properly called 'Mighty Antar' in military service - was a great success, remaining in service in various guises for some 30 or so years and only eventually being replaced in the early 'eighties by the 65 tonne Scammell 'Commander'.

DEVELOPMENT

In April 1949, Thornycroft received the first order for a massive prime mover for the Iraq Petroleum Company, and began work on what was to become the 'Mighty Antar' tractor. A short announcement appeared in the Commercial Motor magazine in July of 1949, where it was stated that the vehicle was named after the Arabian poet, famed for his powers of strength and endurance.

There was no question of multi-million pound, 5-year gestation periods for design and development in those days; the order was placed in April, and design work began immediately. The design team, which incidentally included a young Alan Townsin, now a respected commercial vehicle author, was led by C E Burton, Thornycroft's chief designer. A prototype was running by December of the same year.

In 1950 the vehicle was announced to the trade press, and it was shown later that year at the Earls Court Commercial Motor Show. The 'Antar' was the largest commercial tractor available in Britain at that time, and must have created enormous interest at the time of its launch. There was hardly a copy of the weekly Commercial Motor produced during the years 1950/52 which did not feature a photograph or story of the 'Antar' tractor carrying-off some new and amazing feat of heavy haulage derring-do.

The first commercial 'Antar' tractors were equipped with 250bhp 'Meteorite' Mk 101 indirect-injection diesel

CHAPTER 6.1: THORNYCROFT ANTAR FV12000 SERIES

engines, and were supplied with a skeletal trailer, specially designed by Cranes for carrying 28m lengths of 762mm diameter steel oil-supply pipeline. The vehicle was of 6x4 configuration, with the engine driving twin Kirkstall Forge bogie rear axles via a four-speed main gearbox and three-speed auxiliary box; semi-elliptical springs were used all round. At an all-up train weight of 100 tons, these 'Antars' were expected to haul their 85 ton payloads across some of the world's toughest terrain.

In August 1949, at a meeting called to discuss the post-war range of 'B' vehicles, the Director, Fighting Vehicle Production (DFVP) called the meeting's attention to 'a vehicle being developed by Thornycrofts... called the 'Mighty Antar'... which should tow an 82 ton load'. Expressing keen interest in the vehicle, the Deputy Director, Weapons Development (DDWD) pointed out that the Ministry of Supply had ordered some fifty of the FV3601 50-ton trailer (to transport 'Centurions') with only the 'rapidly-wasting' Diamond T tractors available as prime movers and that there was some urgency to find suitable tractors. It was agreed that the availability and suitability of the 'Antar' should be investigated during 1951/52, with a view to considering it as an alternative prime mover.

The 'Antar' was to be graded GS in the post-war FVRDE vehicle classification scheme, and thus was not strictly a 'cross-country' vehicle, and in bureaucratic terms, could not be considered a direct replacement for the projected FV1000/FV1200 tractors. However, the military authorities showed considerable interest both in the prototype vehicles themselves, and in the trials conducted under various conditions. In 1951, the Ministry of Supply placed the first order (contract 6/Veh/5302/CB27a) for 15 Mk 1 'Antar' tractors modified for military purposes; the major difference being that the diesel 'Meteorite' was replaced by a Mk 204 carburetted unit, redesigned to run on petrol.

No user trials of the Mk 1 'Antar' were undertaken before deliveries were made and this was apparently to lead to subsequent maintenance problems. Although the vehicles were generally reliable in use, it is interesting that during the user trials for the FV3001 semi-trailer, the FVRDE report made the point that the complete rig, consisting of 'Antar' and trailer, 'required highly skilled drivers to operate it' and that due to its size and weight, 'it would be necessary to conduct detailed reconnaissance of routes over which it would have to operate'. Imagine if the FV1000 project had gone ahead!

The Mk 1 vehicle was fitted with a permanent, steel ballast body, and was intended for use with the FV3601 50 ton tank transporter trailer. The trailer drawbar gear could also be used to allow tractors to be 'double-

FV12001 Mk 1 steel ballast-bodied tractor on test at FVRDE (IWM)

Rear view showing stowage and ballast arrangements (IWM)

Early production example of FV12001 Mk 1 tractor (Thornycroft)

CHAPTER 6.1: THORNYCROFT ANTAR FV12000 SERIES

Late version of FV12001 tractor exhibited at FVRDE, 1954 (IWM)

Model of FV12002 Mk 2 fifth-wheel tractor and FV3001 trailer (IWM)

Production example of FV12002 (Thornycroft)

headed' and the braking system airlines could also be coupled together for this purpose.

It wasn't long before the Royal Corps of Transport also decided that a fifth-wheel tractor would be pretty handy, and deliveries of a Mk 2 version began in 1953, initially for this purpose.

The first Mk 2 vehicle was the FV12002, a fifth-wheel tractor for use with the FV3011 50 ton and FV3001/FV3005 60 ton semi-trailers. The Mk 2 differed from the early vehicles only in minor details, for example the fuel tanks were moved from across the chassis to a longitudinal position, and the stowage lockers were repositioned accordingly.

In 1955, FVRDE issued Specification 9173, entitled 'Production specification: tractor, 30 ton, 6x4, GS, Thornycroft Antar Mk 2, FV12003'. With hindsight, this seems a pretty strange thing to do since, by the time the specification was issued, the Mk 1 vehicles had already been in service three or four years, deliveries of Mk 2 vehicles had been underway for some time, and anyway the 'Antar' had already existed in commercial form before the Ministry of Supply placed any orders at all. Presumably bureaucratic necessity demanded that the paperwork remained in order.

However, aside from the body being removable, the major difference between the Mk 1 and Mk 2 ballast tractors was that in the latter case, the ballast body was wooden panelled. It was also possible to convert the Mk 2 vehicles from the fifth-wheel role to the ballast tractor, and back again.

The Mk 1 and Mk 2 tractors remained in service until the early-sixties, but by this time it had become apparent that the relatively-slow speed of the vehicle could be a hazard on the road, and that more power was required. In late 1958, the Mk 3/3A version was submitted for prototype reliability trials. Phase 1 of the trials, which covered some 17,500km, was completed, and a report issued by December of the same year, with only very minor points arising. Deliveries began the following year.

The Mk 3/3A was equipped with a Rolls-Royce C8SFL supercharged, straight-eight diesel engine producing a hefty 333bhp (gross), and a new six-speed transmission system designed to improve economy without sacrificing performance. In fact, the Mk 3/3A 'Antar' was an altogether more-modern machine, and was available configured as a fifth-wheel tractor (FV12004), the Mk 3, and as a ballast tractor (FV12006), Mk 3A.

Although the overall massive appearance was unchanged, the cab was more streamlined with better-shaped front

CHAPTER 6.1: THORNYCROFT ANTAR FV12000 SERIES

and rear wings, and the reduction in width of the twin radiators to a large single radiator allowed the engine compartment covers to be reshaped. Although possibly widened by the insertion of a flat panelled area in the centre, the cab was essentially a standard Bonallack product and was also used on other Thornycroft commercial vehicles; because of this, there was a degree of mismatching of panels around the scuttle and bonnet area.

Like the Mk 2 version, the Mk 3/3A was constructed in both fifth-wheel and ballast tractor configurations, and could be converted from one role to the other. FVRDE Specification 9683, issued in 1962, detailed the body which was used to convert the FV12004 tractor to FV12006 form.

Production

The chassis numbers suggest that a total of 90 Mk 1 and Mk 2 vehicles was supplied under five War Office or Ministry of Supply contracts, and one Air Ministry contract; there were 105 Mk 3/3A vehicles, supplied under two contracts.

'Antars' were produced between 1949 and perhaps 1963, and the total production figure, for both the military and commercial markets, was 750 units.

NOMENCLATURE and VARIATIONS

Aside from the purely civilian versions offered on the commercial vehicle market, the British military 'Antar' was produced in Mk 1, 2 and 3/3A form, and in five major variants, all of them intended primarily as tank transporter tractors. The War Office also took delivery of an additional version (FV12007) which was essentially a commercial tractor design, not intended as a tank transporter.

The 'Antar' military tractor was supplied configured as a ballast tractor, intended for use with full trailers, and as a fifth-wheel tractor, for use with a semi-trailer. The Mk 3/3A 'Antar' was available in 6x4 and 6x6 versions; all other variants were 6x4 only.

All Mk 1 and Mk 2 'Antar' tractors appear to have started life rated at 30 tons, but the weight classification was subsequently upgraded to 50 tons for full trailer tractors, and 60 tons for semi-trailer versions. FVRDE stated that, *under favourable conditions*, the vehicles were capable of operating at a gross train weight of 142,500kg.

Variations

FV12001. Tractor, 30 ton, GS, permanent body, 6x4, Thornycroft Antar, Mk 1
Original Rover 'Meteorite' petrol-engined military 'Antar'; fitted with permanent, steel ballast body intended

Tank transporter rig (FV12002 + FV3001) on test at FVRDE (IWM)

Production example of FV12002 with FV3001 semi-trailer (TMB)

Production example of FV12003 Mk 2 ballast-bodied tractor (BCVM)

CHAPTER 6.1: THORNYCROFT ANTAR FV12000 SERIES

Rear view of same vehicle - note wheel lifting flanges (BCVM)

Tank transporter rig with Centurion (Thornycroft)

FV12004 Mk 3 fifth-wheel tractor (BCVM)

for use with the FV3601 50 ton transporter trailer. Full-width engine compartment incorporating twin radiators; simple flat-panelled cab, initially produced by Bonallack; steel rear body designed for loading with ballast (normally about 15,000kg but occasionally up to 27,500kg); 20 ton, chain-drive winch installed behind the cab.

There were some minor differences in headlamp position, fuel tank layout, and stowage arrangements on early (up to vehicle 7360), as compared to later vehicles.

FV12002. Tractor, 60 ton, GS, for semi-trailer, 6x4, Thornycroft Antar, Mk 2

Fifth-wheel tractor designed for use with the FV3011 50 ton and FV3001/FV3005 60 ton semi-trailers, and intended for tank transporter and heavy road-train duties. Full-width engine compartment and twin radiators; simple, flat-panelled cab by Bonallack or Thornycroft; 20 ton chain-drive winch installed between cab and fifth-wheel platform; fitted with hydraulic equipment and couplings for the trailer ramps and jacks.

There were minor differences in headlamp position, fuel tank layout, and stowage arrangements on early, as compared to later examples.

Could be converted to a ballast tractor by the fitment of a wooden ballast body in which case the designation would become FV12003 (see below).

FV12003. Tractor, 50 ton, GS, ballast body, 6x4, Thornycroft Antar, Mk 2

Ballast tractor intended as a prime mover for the 50 ton FV3601 trailer, and designed for transporting 'Centurion' tanks. Full-width engine compartment and twin radiators; Bonallack or Thornycroft produced cab, and simple, wooden-panelled removable rear ballast body; 20 ton chain-drive winch installed behind cab.

Could be converted to a fifth-wheel tractor by the removal of the ballast body, in which case the designation would become FV12002 (see above).

Also sometimes referred to as Mk 1B.

FV12004. Tractor, 60 ton, GS, for semi-trailer, 6x4, Thornycroft Antar, Mk 3

Development of the FV12002, Mk 2 vehicle, fitted with more-powerful Rolls-Royce C8SFL supercharged diesel engine, new transmission and improved steering. New cab design based on contemporary Thornycroft commercial cab; reduced frontal area with single radiator; 20 ton winch installed beneath small canvas-covered enclosure between cab and fifth-wheel platform. Designed for use with FV3001/FV3005 60 ton semi-trailers, and intended for tank transporter and heavy road-train duties.

CHAPTER 6.1: THORNYCROFT ANTAR FV12000 SERIES

FV12005. Tractor, 50 ton, GS, for semi-trailer, 6x6, Thornycroft Antar, Mk 3
Experimental version of FV12004, Mk 3 vehicle with six-wheel drive.

FV12006. Tractor, 50 ton, GS, for full trailer, 6x4, Thornycroft Antar, Mk 3A
Development of the FV12004, Mk 3 vehicle, fitted with more-powerful Rolls-Royce C8SFL super-charged diesel engine; converted to the ballast tractor role by the addition of a removable steel-framed wooden ballast box, intended for use with the FV3601 50 ton transporter trailer; the fifth wheel of the basic Mk 3 FV12004 was not removed. New cab design based on contemporary Thornycroft commercial vehicles; reduced frontal area with single radiator; 20 ton winch installed beneath small canvas-covered enclosure behind the cab.

FV12007. Tractor, 30 ton, GS, for semi-trailer, 6x4, Thornycroft Antar, Mk 3
Basic commercial tractor, lightly-modified for military applications, and intended for use with a 20 tonne semi-trailer (FV3521).

Modifications

Soon after Thornycroft was absorbed by AEC (or ACV as it had become by then), a 60 ton 'Antar' ballast tractor (FV12002) was equipped with a six-cylinder AEC diesel of 17,750cc and issued for unit assessment. It remained in use from 1963 until 1971 when it was sold.

One of the major problems with the 'Antar' tractor was its weight. The vehicle weighed-in at more than 20,000kg, and with the cancellation of the FV1000 and FV1200 tractors, there was no recovery vehicle able to lift one end of the 'Antar' and recover it on suspended tow. One ingenious solution to this problem was a bolt-on jib, developed by the RASC, which allowed a fifth-wheel 'Antar' tractor to be converted to a recovery vehicle, able to tow a second such tractor.

While the bolt-on jib was considered an 'unofficial', albeit expedient and effective solution, a more long-term answer was provided by the FV2451 10 ton two-wheeled recovery dolly (see Chapter 7).

TRAILERS

The 'Antar' was intended to be used with the following trailers/semi-trailers:

FV12001 ballast tractor:
FV3601. Trailer, 50 ton, 16TW/4LB, transporter, No 1, Mk 3; manufactured by R A Dyson & Co Ltd (chassis), with Crane Fruehauf Trailers Ltd (body), previously known as Cranes (Dereham) Ltd.

Rear view of FV12004 Mk 3 fifth-wheel tractor (BCVM)

FV12004 Antar tractor dwarfs an FV1801 Champ (NMM)

FV12005 Mk 3 ballast-bodied tractor (KP)

CHAPTER 6.1: THORNYCROFT ANTAR FV12000 SERIES

Beautiful factory shot of FV12005 before delivery (Thornycroft)

FV12005 tractor in use - note missing fuel tank guard (NMM)

Line-up of FV12005 tractors awaiting completion at Basingstoke (BCVM)

FV12002 fifth-wheel tractor:
FV3001. Semi-trailer, 60 ton, 8TW/2LB, tank transporter (road); manufactured by Joseph Sankey & Sons Ltd.
FV3005. Semi-trailer, 60 ton, 8TW/2LB, tank transporter (road); manufactured by Joseph Sankey & Sons Ltd. Improved version of FV3001.
FV3011. Semi-trailer, 50 ton, 8TW/2LB, tank transporter (road); manufactured by Joseph Sankey & Sons Ltd, Tasker Trailers Ltd, and GKN Group Services Ltd.

FV12003 ballast tractor:
FV3601. Trailer, 50 ton, 16TW/4LB, transporter, No 1, Mk 3; manufactured by R A Dyson & Co Ltd (chassis), with Crane Fruehauf Trailers Ltd (body), (previously known as Cranes (Dereham) Ltd).

FV12004 fifth-wheel tractor:
FV3001. Semi-trailer, 60 ton, 8TW/2LB, tank transporter (road); manufactured by Joseph Sankey & Sons Ltd.
FV3005. Semi-trailer, 60 ton, 8TW/2LB, tank transporter (road); manufactured by Joseph Sankey & Sons Ltd. Improved version of FV3001.
FV3011. Semi-trailer, 50 ton, 8TW/2LB, tank transporter (road); manufactured by Joseph Sankey & Sons Ltd, Tasker Trailers Ltd, and GKN Group Services Ltd.

FV12006 ballast tractor:
FV3601. Trailer, 50 ton, 16TW/4LB, transporter, No 1, Mk 3; manufactured by R A Dyson & Co Ltd (chassis), with Crane Fruehauf Trailers Ltd (body), previously known as Cranes (Dereham) Ltd.

FV12007 fifth-wheel tractor:
FV3521. Semi-trailer, 20 ton, 4TW/2LB, low loading; manufactured by Taskers of Andover (1932) Ltd, Hands (Letchworth) Ltd, and British Trailers & Co Ltd.

DESCRIPTION

Engine

Mk 1 and Mk 2 vehicles
As with the prototype FV1000 and FV1200 vehicles, FVRDE would have preferred that the 'Antar' be fitted with a diesel engine, but the diesel version of the 'Meteorite' used by Thornycroft on the commercial tractors was not felt to be sufficiently powerful. No other suitable diesel power units were available at the time, and once again, the Rolls-Royce designed, Rover 'Meteorite' V8 petrol engine was used as a stop-gap measure.

Unlike the engines installed in the two purpose-designed CT tractors, the Mk 204 version of the 'Meteorite' selected for the military 'Antar' was a normally-aspirated unit, breathing through a couple of monstrous Solex 46 ZNHP carburettors. The 60° angle of the cylinder blocks was not ideal for the V8 configuration and this apparently resulted in a rather uneven idle and some roughness through the range.

With a compression ratio of 6:1, the Mk 204 had a gross power output of 260bhp, while the Mk 202 employed in the FV1000/FV1200 produced almost 500bhp (gross) from a compression ratio of 7:1. It may well be that the use of carburettors, combined with a lower compression ratio allowed the 'Antar' to operate on 70 octane fuel, where the other tractors required 80 octane. However, as can be seen there was a considerable sacrifice in gross power output; it is also interesting to note that the maximum rev limit was reduced when compared to the direct-injected versions, and that the maximum torque produced was also significantly lower.

The engine consisted of a pair of aluminium-alloy cast crankcases (metal to Air Ministry specification DTD 133), each fitted with an aluminium cylinder head, with overhead inlet and exhaust valves. There were four valves per cylinder, operated by overhead cams (the diesel version fitted to the commercial 'Antars' used only three valves per cylinder). Fuel was supplied to the twin carburettors by a pair of David 'Korrect' mechanical diaphragm-type fuel pumps, one mounted on each block and driven by the camshafts. The fuel consumption was in the order of 2.25 litre/km giving a total range from the two 450 litre tanks of 400km.

The lubrication system was of the dry-sump type, with a total oil capacity of 55 litres; the oil cooler tubes for the engine lubrication system were mounted ahead of the right-hand radiator.

Each cylinder was provided with a pair of screened spark plugs, rather confusingly described as 'inlet' or 'exhaust' according to which side of the cylinder block they were mounted. Although a booster coil was fitted for starting, the ignition was by a twin-magneto system, employing a pair of FVRDE-designed screened magnetos, designated 'No 1, Mk 1' on the right-hand cylinder bank, and 'No 1, Mk 2' on the left-hand bank; early engines employed British Thompson Houston C8B/1A left-hand and right-hand magnetos. The ignition system wiring harness was designed to allow the engine to continue to function normally with one magneto defective.

Fuel was supplied to the engine through a pair of Solex 46 ZNHP carburettors.

Mk 3/3A vehicles
By the time the Mk 3/3A vehicle was introduced, a suitable diesel engine was available and FVRDE were at last able to achieve their original goal. All of the Mk 3/3A (military) 'Antars' were fitted with the Rolls-Royce C8SFL-843 power unit. Apparently Rolls-Royce were reluctant to allow this unit to be used in the later commercial 'Sandmaster' models, and these had to make do with a turbo-charged six-cylinder C6TFL.

The C8SFL was a water-cooled, straight-eight supercharged diesel engine, with a capacity of 16.2 litres, and fitted with a Roots-type blower. Engine speed was governed to 2100rpm giving a gross power output of 333bhp.

The engine consisted of a monobloc casting, with push-fit wet cylinder liners; the pistons were provided with open-cavity combustion chambers in their crowns giving a 14:1 compression ration (compared to 16:1 on the normally-aspirated versions). The valve configuration was overhead, with two valves per cylinder, operated by conventional pushrods operated by a low-mounted camshaft. Valve seats and stem tips were Stellite faced to prevent wear, and the exhaust valves were sodium-filled to assist with cooling.

Engine data

	Meteorite Mk 204	C8SFL-843
Configuration	V8	Straight 8
Fuel	70 octane petrol	DERV
Capacity (cc)	18,012	16,200
Bore and stroke (in)	5.40 x 6.0	5.125 x 6.0
Compression ratio	6:1	14:1
Firing order:*		
chassis nos 84001-6	A1 B4 A3 B2 A4 B1 A2 B3	-
chassis nos 84007 up	A1 B2 A3 B1 A4 B3 A2 B4	-
Mk 3/3A vehicles	-	16258374
Power output (bhp):		
gross	260	333
net	250	313
Engine speed (rpm):		
maximum	2300	2250
governed	2000	2100
Maximum torque (lbs/ft)	728	934

* Offside cylinder block designated 'A', nearside block 'B'.

Cooling system
Both types of engine were water cooled, with the cooling system pressurised to 7050kg/m² on the Mk 1 and 2 vehicles, and 3524kg/m² on Mk 3/3A vehicles. The capacity of the cooling system on the twin-radiator versions was 110 litres; 52 litres on the Rolls-Royce engined Mk 3/3A vehicles.

The cooling water was passed through a pair of parallel-flow Morris block radiators (one only on Mk 3 vehicles); a belt-driven cooling fan was mounted in between the engine and each of the radiators.

Transmission

The drive was transmitted from the engine via a short propeller shaft to a 455mm diameter Borg and Beck No 18 R4 twin dry-plate clutch mounted remote from the engine, and in unit with the main and auxiliary gearboxes. The clutch-release mechanism was power-assisted from the braking system air-pressure reservoir, and an air-pressure operated mechanical clutch-brake system was provided to speed the gear changing. On Mk 1 and Mk 2 vehicles, a 13-blade cooling fan was incorporated in the clutch housing for ventilation purposes, together with a ring-and-pinion gear to allow the engine to be turned over by hand (presumably for ignition timing or other mechanical work, rather than for starting it!).

There were fundamental differences in the transmission according to the vehicle 'mark'.

Mk 1 and Mk 2 vehicles were fitted with a constant-mesh main gearbox, with dog engagement, giving four forward speeds and one reverse. A three-speed auxiliary gearbox was provided, with ratios of 0.732:1, 1:1, and 1.728:1. The main gearbox also incorporated a power take-off, driven at 0.62 times engine speed, for the winch, and a separate drive was provided for the braking system compressor which was mounted on the gearbox casing.

Mk 3 vehicles received a completely-redesigned transmission system with a combined main and auxiliary gearbox. Some of the 12 available speeds (four-speed gearbox x three-speed auxiliary) were not considered useful, so the auxiliary gearbox was removed, and the main gearbox modified to give six forward speeds, with fifth gear being direct, and sixth an overdrive speed. The gearbox was also of the constant-mesh type with engagement by sliding dogs.

A hydraulic pump, mounted on the main gearbox was designed to provide hydraulic power for the steering system; on vehicles intended for use with semi-trailers, a second hydraulic system was fitted to operate the ramps and jacks on the semi-trailer.

Final drive from the auxiliary gearbox to the forward of the two rear axles was by means of an open Hardy-Spicer propeller shaft; a short, inter-axle shaft transmitted the drive to the second axle.

Axles and suspension

The vehicle used live axles at both front and rear, suspended on semi-elliptical springs; there were no shock absorbers. The front axle was a solid forged rectangular section, with a swivel eye at each end; each rear axle housed a worm wheel and epicyclic reduction gear.

The front springs were attached to pins at each spring eye, with rear shackle plates, while the massive (125mm wide x 2175mm long) rear springs were pivoted at the centre by means of bolted spring boxes. The rear axles were supplied by Kirkstall Forge Engineering, and were located by ball-ended radius arms designed to handle the forces generated in the bogie by braking and slewing. These ball-ended arms allowed bogie articulation for diagonally-opposite wheels of 380mm.

Front suspension movement was restricted by the use of rubber bump stops, while movement of the rear axle bogie was restricted by slings attached to the bogie itself.

A third differential was installed between the two rear axles on Mk 3/3A vehicles, and was lockable manually.

Steering gear

A power-assisted steering system was fitted, consisting of Marles spiral cam-and-roller mechanism with hydraulic-ram assistance; on Mk 1 and Mk 2 vehicles, there was apparently little assistance when the engine was running at less than 1200rpm. The wheel required 5.75 turns from one lock to the other to give a turning circle of nearly 20m. On Mk 3/3A vehicles, the steering wheel was adjustable within a 75mm range.

The steering arrangements actually consisted of two separate steering systems; one connected directly from the steering box to the right-hand stub axle steering arm, the other connected by means of two hydraulic piston assemblies to the left-hand stub axle. The two stub axles were interconnected by means of a conventional tie rod.

Minor improvements were made to the power steering system on Mk 3/3A vehicles.

Braking system

Air-pressure brakes were provided, operating on all three axles; a twin-cylinder compressor was installed on the side of the main gearbox, driven by the layshaft, or on Mk 3/3A vehicles installed on the engine and driven by V belts. A methanol anti-freeze system was provided to prevent moisture freezing in the brake lines.

Air-line connections were installed at the front and rear of the vehicle to allow tandem operation under a single control, and to permit the trailer brakes to be connected to, and operated by the tractor. A hand reaction valve was also provided in the cab to permit independent operation of the trailer brakes, for example during descent of steep hills.

The brake drums were 483mm diameter x 102mm width at the front, and 178mm at the rear.

The handbrake was operated through a mechanical linkage to the bogie brakes; the handbrake was also interconnected with the air pressure system to provide additional power-assisted operation of the bogie brakes whenever air pressure was available.

On Mk 3/3A vehicles, there was also a Clayton 'Oetiker' exhaust brake, designed to provide braking by engine back pressure with the exhaust pipes closed off. The brake was operated by means of a small control lever installed just forward of the gear lever.

Road wheels
The vehicle was fitted with single front wheels, and dual rears. Lifting flanges were supplied with each vehicle for attachment to the front and rear hubs, allowing the vehicle to be easily hoisted from the wheels, for example onto the deck of a ship (remember, this was in the days when the only 'ro-ro' ferries were military landing craft). It was suggested that the vehicle should not be operated with the flanges in position.

The wheels were manufactured by Dunlop, and were of the four-piece disc type, size 10.00x24in; fitted with Goodyear 'Hard Rock Grip', Dunlop 'Power Grip' or Firestone 'Rock Grip' tyres, size 14.00x20.

A spare wheel was carried, either behind the cab or in the ballast box. On the ballast-bodied tractors, a crane-and-davit system was provided for handling the spare wheel. The brake system compressor could also be used for tyre inflation.

Chassis
The frame was of bolted construction, consisting of 292x89x9.5mm channel-section main members, fitted with full-length 6.4mm thick liners, and with additional 225x9.5mm stiffening plates installed each side at the rear bogie. Main cross members were formed by back-to-back channel-section members, diagonally braced with angle sections and gusset plates.

The engine was mounted on a rubber-isolated channel-section sub-frame, with torque arms either side, designed to carry the engine torque stresses to the chassis.

Fifth-wheel equipment
The fifth-wheel equipment, designed by FVRDE (FV252261) was of conventional pattern, with a manual lock for the trailer kingpin; twin ramps were installed behind the turntable, to help lift and guide the trailer into position as the tractor backed up towards it.

All Mk 3/3A vehicles were fitted with a fifth wheel, regardless of the body type.

Cab and bodywork
Cab
On Mk 1 and Mk 2 vehicles, the simple, flat-panelled three-man cab was initially supplied by Bonallack, but eventually produced in-house by Thornycroft to the same design using panels supplied by Motor Panels Ltd. The cab was of steel-framed, twin-skin construction, with the cavity filled with 'Isoflex' insulating material, and was attached to the chassis by means of a flexible four-point mounting system. The floor was of steel sheet, uninsulated, but provided with a rubber mat covering.

Still of twin-skin insulated design, the cab used on Mk 3/3A vehicles was also supplied by Bonallack, but was of a more-modern pressed-steel type using compound-curved panels.

In both cases, adjustable ventilators were provided on each side of the cab to provide an adequate air flow, and the roof was fitted with a circular observation hatch and anti-aircraft hip ring above the passenger's seat.

A fixed, laminated-glass two-piece windscreen was fitted, with drop-down lights and swivelling quarter-lights in the cab doors.

There were two seats in the cab: a small, adjustable bucket seat for the driver, and a fixed bench seat for the two passengers. The base of the passenger seat also served to house the batteries.

Rear body
On tractors intended for semi-trailers, there were two large stowage compartments, one installed either side of the winch.

The ballast bodies were sectionalised to house cast-iron weights, with the wooden composite ballast bodies supplied as a temporary fitment only. A small compartment was provided in the body to house the spare wheel, and a detachable davit was carried to enable the heavy wheel and tyre to be handled. Large sheet-steel mudflaps were fitted behind the rearmost axle, and ahead of the centre axle on steel-bodied vehicles.

On Mk 2 and Mk 3 fifth-wheel tractors, a tubular-steel framework was often installed behind the cab, with a covering tarpaulin, intended to provide a housing for the winch and accommodation for the crew.

Electrical system
The vehicles were wired on a 24V negative-earth system.

On Mk 1 and Mk 2 vehicles, there were either four 6V 110Ah batteries, or 18 separate 160Ah cells; on Mk 3/3A vehicles, eight 150Ah batteries were used, connected in series/parallel. The contemporary documentation suggests that vehicles supplied to the RAF under contract

5718 were fitted with six nickel-cadmium alkaline (NIFE) batteries, each battery consisting of three cells.

With the exception of the tail lamps, which for some reason used a conventional earth return, all components were connected using double-pole wiring.

A CAV axial-type starter was fitted to Mk 1 and Mk 2 vehicles; with a single-speed CAV 175mm diameter generator, giving a maximum output of 33A at 750rpm. Mk 3/3A vehicles were fitted with twin CAV MS624/W4 starters, with a 55A output CAV generator.

Winch

A 20 ton capacity mechanical, horizontal-drum winch was installed behind the cab, chain-driven by a power take-off from the main gearbox; the winch was a Darlington type 70/31/P26 on Mk 1 and Mk 2 vehicles, and a Turner on Mk 3/3A vehicles. All of the winch controls were accessible from inside the cab, and on Mk 3/3A vehicles, the winch itself was provided with a small canvas-covered enclosure.

The Darlington winch employed 22.2mm diameter rope, with a total length of 107m. With the engine running at 1000rpm, the maximum winch speed was 23.5m/min winching in, and 17m/min winching out. Rope diameter on the Turner winch was 25mm, with a total length of 111m; maximum speeds was 15m/min at 1000rpm engine speed in fourth gear. Both types of winch were fitted with a rope overload device, this took the form of a shear pin on the Darlington machine, and an electrical cutout on the Turner.

Guide rollers and pulleys were provided to allow the winch rope to provide rear pulls at angles of up to 40° above and below the horizontal, and 20° from either side. The winch was provided for the purpose of loading disabled AFV's to the vehicle trailer, rather than for self-recovery, and thus no provision was made for winching from the front of the vehicle.

DOCUMENTATION

Technical publications

Specifications
FVRDE Specification 9173. Production specification: tractor, 30 ton, 6x4, GS, Thornycroft Antar Mk 2, FV12003.
FVRDE Specification 9207. Production specification: tractor, 30 ton, 6x4, GS, Thornycroft Antar Mk 3, FV12004.
FVRDE Specification 9683. Production specification: body and mounting for tractor, 30 ton, 6x4, GS, Thornycroft Antar Mk 3 for semi-trailer, FV12004.

Reports
FVRDE Report FT/B 345. Interim report on semi-trailer, 60 ton, 8TW/2LB, tank transporter, FV3001.
FVRDE Report FT/B 577. Development trials: tractor, 30 ton, 6x4, GS, Thornycroft Antar, Mk 3, FV12004; Phase 1.

User handbooks
Provisional user handbook: tractor, super heavy, 6x4, Thornycroft Mighty Antar. Air publication 4357A (UH).

User handbook: tractor, 60 ton, GS, 6x4, for semi-trailer, Thornycroft Antar; tractor, 50 ton, GS, 6x4, Thornycroft Antar. WO codes 17769, 18414.

Servicing schedules
Tractor, 60 ton GS, 6x4, for semi-trailer, Thornycroft Antar; tractor, 50 ton, GS, 6x4, Thornycroft Antar. WO code 11071, 13057; Army codes 60057, 60057/1, 60057/2, 60058.
Tractor, 60 ton, GS, for semi-trailer, Thornycroft Antar, Mk 3; tractor, 60 ton, GS, 6x4, Thornycroft Antar, Mk 3A. WO code 13898.

Parts lists
Tractor, 6x4, super heavy, Mighty Antar. WO codes 17858, 17981.
Tractor, 30 ton, GS, 6x4, for semi-trailer, Mighty Antar, Mk 2. Army code 19996.
Tractor, 50/60 ton, GS, 6x4, for semi-trailer, Thornycroft Antar, Mk 3/3A. Army code 20832.

Engine, Meteorite, Mk 204, installed in tractor, GS, super-heavy, 6x4, Thornycroft Mighty Antar. WO codes 17708, 12691.

Technical handbooks
Data summary:
Tractor, 60 ton, GS, for semi-trailer, Thornycroft Antar; tractor, 50 ton, GS, 6x4, Thornycroft Antar. EMER R610.
Tractor, 60 ton, GS, for semi-trailer, Thornycroft Antar, Mk 3; tractor, 60 ton, GS, 6x4, Thornycroft Antar, Mk 3A. EMER R620.

Technical description:
Tractor, 60 ton, GS, for semi-trailer, Thornycroft Antar; tractor 50 ton, GS, 6x4, Thornycroft Antar. EMER R612.
Tractor, 60 ton, GS, for semi-trailer, Thornycroft Antar, Mk 3; tractor, 60 ton, GS, 6x4, Thornycroft Antar, Mk 3A. EMER R622.

Engines, diesel, Rolls-Royce C Series. EMER S532.

Repair manuals
Unit repairs:
Tractor, 60 ton, GS, for semi-trailer, Thornycroft Antar; tractor, 50 ton, GS, 6x4, Thornycroft Antar. EMER R613.
Tractor, 60 ton, GS, for semi-trailer, Thornycroft Antar, Mk 3; tractor, 50 ton, GS, 6x4, Thornycroft Antar, Mk 3A. EMER R623.

Engines, diesel, Rolls-Royce C Series. EMER S533.

Field repairs:
Tractor, 60 ton, GS, for semi-trailer, Thornycroft Antar; tractor, 50 ton, GS, 6x4, Thornycroft Antar. EMER R614.
Tractor, 60 ton, GS, for semi-trailer, Thornycroft Antar, Mk 3; tractor, 50 ton, GS, 6x4, Thornycroft Antar, Mk 3A. EMER R624.

Base repairs:
Tractor, 60 ton, GS, for semi-trailer, Thornycroft Antar; tractor, 50 ton, GS, 6x4, Thornycroft Antar. EMER R614 Part 2.
Tractor, 60 ton, GS, for semi-trailer, Thornycroft Antar, Mk 3; tractor, 50 ton, GS, 6x4, Thornycroft Antar, Mk 3A. EMER R614 Part 2.

Field and base repairs: engines, diesel, Rolls-Royce C Series. EMER S534.

Modification instructions
Tractor, 60 ton, GS, for semi-trailer, Thornycroft Antar; tractor, 50 ton, GS, 6x4, Thornycroft Antar. EMER R617, instructions 1-30.
Tractor, 60 ton, GS, for semi-trailer, Thornycroft Antar, Mk 3; tractor, 50 ton, GS, 6x4, Thornycroft Antar, Mk 3A. EMER R627, instructions 1-26.
Engines, diesel, Rolls-Royce C Series. EMER S537, instructions 1-5.

Standards
Inspection standard:
Tractor, 60 ton, GS, for semi-trailer, Thornycroft Antar; tractor, 50 ton, GS, 6x4, Thornycroft Antar. EMER R618, Part 1.
Tractor, 60 ton, GS, for semi-trailer, Thornycroft Antar, Mk 3; tractor, 50 ton, GS, 6x4, Thornycroft Antar, Mk 3A. EMER R628, Part 1.

Base inspection standard:
Tractor, 60 ton, GS, for semi-trailer, Thornycroft Antar; tractor, 50 ton, GS, 6x4, Thornycroft Antar. EMER R618, Part 2.
Tractor, 60 ton, GS, for semi-trailer, Thornycroft Antar; tractor, 50 ton, GS, 6x4, Thornycroft Antar. EMER R628, Part 2.
Engines, diesel, Rolls-Royce C Series. EMER S538, Part 2.

Miscellaneous instructions
Tractor, 60 ton, GS, for semi-trailer, Thornycroft Antar; tractor, 50 ton, GS, 6x4, Thornycroft Antar. EMER R619, instructions 1-9.
Tractor, 60 ton, GS, for semi-trailer, Thornycroft Antar, Mk 3; tractor, 50 ton, GS, 6x4, Thornycroft Antar, Mk 3A. EMER R629, instructions 1-13.

Engines, diesel, Rolls-Royce C Series. EMER S539, instructions 1-9.

Complete equipment schedule
Tractor, 60 ton, GS, for semi-trailer, Thornycroft Antar, Mk 3; tractor, 60 ton, GS, 6x4, Thornycroft Antar, Mk 3A. WO code 33849.

Tools and equipment
Tables of tools and equipment for 'B' vehicles: tractor, 60 ton GS, 6x4, for semi-trailer, Thornycroft Antar, WO code 18496, Table 1093; tractor, 50 ton, GS, 6x4, Thornycroft Antar, WO code 17769, Table 1315.

Bibliography
Antar antics. London, Commercial Motor magazine, 10 March 1950.

Breakdown. A history of recovery vehicles in the British Army. Baxter, Brian S. London, HMSO, 1989. ISBN 0-11-290456-4.

Britain's most powerful vehicle. London, Commercial Motor magazine, 3 March 1950.

The illustrated history of Thornycroft trucks and buses. Baldwin, Nick. Sparkford, Haynes Publishing Group, 1989. ISBN 0-85429-707-3.

TRAILERS and SEMI-TRAILERS
CHAPTER 7

CHAPTER 7: TRAILERS and SEMI-TRAILERS

CHAPTER 7.1

TRAILERS and SEMI-TRAILERS
Rogers M9; 40 ton Mk 1/Mk 2; FV2751, FV3001/FV3005
FV3011, FV3221, FV3541, FV3551, FV3561, FV3601, FV3621, FV35003

In recent years, moving tanks, or for that matter any heavy machinery on roads, has become less of a problem. Improvements in the power output of diesel engines have generally been matched by similar advances in transmission and braking systems, and in tyre and suspension technology. A modern commercial tractor such as a Volvo F10, for example, is perfectly capable of hauling 100 tons on the highway at quite reasonable speeds, and aside from reliability and durability issues, there is probably no need for purpose-designed equipment for road use... of course, the military authorities still persist with expensive, specialised tractors, but that's another story.

Fifty years ago, things were not quite so straightforward. Big truck technology was in its infancy, and those trucks which did exist had almost all been designed by, or on behalf of the military, and whenever off-road performance was required, were designed without compromise... which tended to mean to a high price. While diesel engines were nowhere near as flexible or powerful as they are today, the larger petrol engines tended to be complex, expensive and thirsty. And of course, once suitable tractors were developed, there was the problem of actually loading and carrying the tank, or other item of equipment. To make matters worse, where a measure of off-road performance was required from the tractor, then obviously the trailer was also required to provide the same level of performance.

It was not long before two different approaches to the load-carrying problem emerged: the semi-trailer for use with the fifth-wheel tractor, and the full, or draw-bar trailer, designed to be towed by a ballast tractor or so-called prime mover. Each system had its own advantages and disadvantages.

One constant problem with all of the larger trailers designed to carry substantial loads, regardless of configuration, was that of tyre blow-outs, and for this reason, puncture detector systems were often employed. Inevitably, with these multi-wheel trailers, it would be an inner wheel which picked up the first puncture. If this was not attended to immediately, the increased loading on the other tyres would soon lead to further tyre failures. Perhaps it was cause, perhaps it was effect, but the transporter crews would often choose not to deal with single punctures, but would wait for the inevitable second or third blowout before, literally being forced to stop.

TUGS OF WAR

157

CHAPTER 7.1: TRAILERS and SEMI-TRAILERS

This chapter includes full trailers for use with ballast tractors, and semi-trailers of both types. Among the semi-trailers, those intended for use with the Albion CX22S, the Scammell 'Pioneer', the Diamond T, and the prototype FV1000 were 'dedicated' to the specific tractor; these are dealt with in the respective chapters dealing with the vehicles in question. However, by the time the Scammell 'Constructor' and Thornycroft 'Antar' were produced, the standardised semi-trailer was well-established, and these tractors could be used with one of a number of trailers.

As well as covering these standardised trailers, this chapter also includes the so-called 'dolly' trailer or dummy axle, which was designed to allow one 'Antar' tractor to haul another on suspended tow.

APPLICATIONS

Trailer	Suitable tractors
Full trailers:	
M9	Diamond T, 980/981
Mks 1/2	Diamond T, 980/981
FV3221	Leyland 'Martian', FV1119
	Scammell 'Pioneer', SV/2S
	Scammell 'Explorer', FV11301
FV3551	Scammell 'Constructor', FV12105
FV3601	Diamond T, 980/981
	Scammell 'Constructor', FV12105
	Thornycroft 'Antar', FV12001/12003/12006
FV3621	Scammell 'Constructor', FV12101/12105
FV35003	Scammell 'Constructor', FV12101
Semi-trailers:	
FV3001	Thornycroft 'Antar', FV12002/12004
FV3005	Thornycroft 'Antar', FV12002/12004
FV3011	Thornycroft 'Antar', FV12002/12004
FV3541	Scammell 'Constructor', FV12102/12105
Dummy axle and dolly trailers:	
FV3561	Thornycroft 'Antar', FV12001/12003/12006
FV2751	Scammell 'Constructor', FV12101

FULL TRAILERS

The full trailer was designed for use with a ballast tractor. This had the advantage that it was not necessary for the tractor to be dedicated to one type of trailer, but it was inevitably more expensive and complex than the semi-trailer.

With a full trailer, the load tends to be spread over a greater number of axles and thus the load-carrying ability is maximised, but on the down-side, since the weight of the trailer is not over the rear wheels of the tractor, ballast weight must also be carried to gain maximum traction. The weight of the ballast reduces the total useful load-carrying ability of the outfit... and

American M9 45 ton trailer hauled by Diamond T (IWM)

British Mk 1/Mk 2 40 ton trailer loaded with a Churchill (REME)

FV3221 10 ton light recovery trailer carrying AEC Militant (IWM)

CHAPTER 7.1: TRAILERS and SEMI-TRAILERS

FV3551 20 ton aircraft fuselage transporter (TMB)

Well-loaded FV3601 50 ton trailer carrying an M26 Pershing (IWM)

FV3621 20 ton trailer for engineering plant (IWM)

because of the tractor-and-trailer configuration, the outfit tends to be rather lengthy.

Steering on a full trailer was normally effected by means of a simple swivelling turntable (non-Ackerman design), fitted between the trailer chassis and the front axle. The trailer would follow the tractor into turns, with the pivot point and radius of turn depending on the position of the draw-bar coupling.

M9 45 ton tank transporter trailer
This was an American-designed and produced three-axle draw-bar trailer produced during WW2, and intended for use as a tank transporter generally in conjunction with the Diamond T 980/981 ballast tractors.

The frame was of all-welded construction, with the bed parallel to the road surface. Loading was from the rear, by means of two triangular folding ramps. Early in the war, the trailer was produced in two types: transporter and recovery. The transporter variant was provided with chock blocks, lashing eyes, and winch cable fairlead rollers; the recovery type was also provided with one anchored, and one movable snatch block. Later trailers were all of the recovery type.

The absence of runway outer guide rails made the trailer particularly useful because it meant that the trailer could carry tanks which exceeded the trailer width.

Two spare wheels were carried under the trailer bed.

Manufacturers
Rogers Inc; Winter Weiss Inc; and possibly others.

Nomenclature
M9. Trailer, 45 ton, 12TW/2LB, transporter.

Dimensions and weight
Length: 9195mm.
Width: 2896mm.
Height: 1448mm.
Track: outer, 2184mm; inner, 635mm.
Weight: laden, 56,101kg; unladen, 10,283kg.
Capacity: 45,818kg.

Suspension and axles
Semi-elliptical springs on the front axle; unsprung walking beams at the rear; four sets of twin wheels on each of three axles.

Braking system
Twin air-line pressure braking system designed to be coupled to the tractor; mechanical, wheel-type hand brake, operating on rear bogie wheels only.

Road wheels
One-piece disc wheels, 7.00x15in, mounting standard road-pattern tyres.

CHAPTER 7.1: TRAILERS and SEMI-TRAILERS

40 ton tank transporter trailer, Mks 1 and 2

The WW2 British 40 ton tank-transporter trailers were similar to the American M9 models described above. Produced in both Mk 1, and improved Mk 2 versions, the British trailers were of the three-axle draw-bar type, and were designed for use as a tank transporter, again normally with the Diamond T 980/981 ballast tractors.

The frame was of all-welded construction, with the bed running more-or-less parallel to the road surface, and with triangular loading ramps provided at the rear. Chock blocks, and winch cable fairlead rollers were provided to assist with loading and securing the payload.

On the Mk 2 trailer, as with the American Rogers design, there were no outer guide rails, which allowed oversize loads to be carried; the inner guide rails were adjustable.

Two spare wheels were carried under the trailer bed.

Manufacturers
Cranes (Dereham) Ltd; R A Dyson & Co Ltd.

Nomenclature
Trailer, 40 ton, 12TW/2LB, transporter, Mks 1 and 2.

Dimensions and weight
Length: 9868mm.
Width: 3050mm.
Height: 1740.
Track: outer, 2400mm; inner, 800mm.
Weight: laden, 54,575kg; unladen, 13,847kg.
Capacity: 40,727kg.

Suspension and axles
Semi-elliptical springs on the front axle; unsprung walking beams at the rear; four sets of twin wheels on each of three axles.

Braking system
Twin air-line pressure braking system designed to be coupled to the tractor; mechanical, wheel-type hand brake, operating on rear bogie wheels only.

Road wheels
One-piece disc wheels, 7.00x15in, mounting standard road-pattern tyres.

FV3221 10 ton light recovery trailer

Simple, flat-bed tandem-axle, eight-wheeled draw-bar trailer designed for use by REME, for the recovery and transportation of wheeled and tracked vehicles up to 10 tons in weight. Normally towed behind the FV1119 Leyland 'Martian', but might also be used with the Scammell 'Pioneer' or 'Explorer' recovery tractors.

Loading was from the rear, by means of loading ramps, normally carried on the trailer deck. Adjustable prop stands were provided at the rear to prevent undue loads

FV35003 10 ton trailer for engineering plant (TMB)

FV3001 60 ton semi-trailer loaded with a Conqueror (IWM)

FV3011 50 ton semi-trailer loaded with a Centurion (TMB)

FV3541 30 ton semi-trailer for engineering plant (TMB)

FV3561 10/30 tonne dummy axle trailer (REME)

FV2751 10 ton dolly trailer converter (TMB)

being placed on the rear suspension during loading and unloading operations.

A standard draw-bar was provided, and there were winch rope fairleads, front and rear to facilitate loading; lashing chains and blocks were provided to secure the load.

Manufacturers
J Brockhouse & Co Ltd; Crossley Motors Ltd; Rubery Owen & Co Ltd.

Nomenclature
FV3221. Trailer, 10 ton, 4TW/2LB, recovery.

Dimensions and weight
Length: 7280mm (excluding draw bar).
Width: 2743mm.
Height: 1715-1900mm, according to manufacturer.
Wheelbase: 4267mm.
Track: outer, 2060mm.
Weight: laden, 18,160kg, emergency maximum, 20,340kg; unladen, 7130kg.
Capacity: 10,181kg.

Suspension and axles
Semi-elliptical springs, front and rear; two sets of twin wheels on each axle.

Braking system
Twin air-line pressure braking system designed to be coupled to the tractor; mechanical hand brake, operating on rear wheels only.

Road wheels
One-piece disc wheels, 7.33x20in, mounting standard road-pattern tyres.

FV3551 20 ton aircraft fuselage transporter
Flat-bed tandem-axle, eight-wheeled draw-bar trailer designed for use by the RAF, for transporting aircraft fuselage up to 20 tons in weight. Normally towed behind the FV12105 Scammell 'Constructor' ballast tractor, and suitable for both on- and off-road use.

Loading would have been by crane, with a special cradle on the trailer bed designed to provide support to the fuselage. The trailer was supplied with a choice of draw bar lengths (3050mm and 8235mm) to cater for different load overhangs, and to simplify operation.

Manufacturers
Taskers of Andover (1932) Ltd.

Nomenclature
FV3551. Trailer, 20 ton, 8TW/2LB, aircraft fuselage transporter.

Dimensions and weight
Length: 12,802mm.
Width: 3505mm.

Height: 1270mm.
Wheelbase: 9830mm.
Track: outer, 3200mm.
Weight: laden, 31,980kg; unladen, 11,650kg.
Capacity: 20,364kg.

Suspension and axles
Semi-elliptical springs, front and rear; four sets of twin wheels on each axle. Both axles were designed to steer, but were not interconnected; this allowed the trailer to be shunted sideways when required.

Braking system
Twin air-line pressure braking system designed to be coupled to the tractor.

Road wheels
One-piece disc wheels, 7.50x15in, mounting standard road-pattern tyres.

FV3601 50 ton tank transporter trailer
Massive 50 ton draw-bar trailer intended for transporting tanks and AFV's up to 50 tons in weight, and normally towed behind all marks of the Thornycroft 'Antar' ballast tractor. Later models were uprated (or perhaps re-rated might be a better description) to 50 tonnes. Supported on four axles, with four sets of twin wheels on each.

The Mk 1 trailers were fully decked to also allow the carrying of other types of equipment. Loading was from the rear by means of four folding ramps. Winch rope fairleads were provided at front on all marks, and at both front and rear on Mk 2 and 3 trailers, to allow disabled loads to be winched aboard; lashing chains and turnbuckles were provided to enable the load to be secured.

Manufacturers
Chassis, R A Dyson & Co Ltd; body, Cranes (Dereham) Ltd, later Crane Fruehauf Trailers Ltd.

Nomenclature
FV3601. Trailer, 50 ton, tank transporter, 16TW/2LB, No 1, Mks 1, 2 and 3.

Dimensions and weight
Length: 10,210mm.
Width: 3200mm.
Height: Mk 1 trailers, 2540mm; Mk 2 and 3 trailers, 2210mm.
Wheelbase: 4570mm
Track: outer, 2400mm; inner, 790mm.
Weight: laden, 69,446kg; unladen, 18,660kg.
Capacity: 50,786kg.

Suspension and axles
Unsprung walking-beams at front and rear; four sets of twin wheels on each of four axles.

Braking system
Twin air-line pressure braking system designed to be coupled to the tractor; mechanical hand brake, operating on rear wheels only.

Road wheels
Three-piece disc wheels, 10.00x20in, mounting standard road-pattern tyres.

Provision was made for carrying either one, or two spare wheels, with a manual winch to simplify handling.

FV3621 20 ton trailer for engineering plant
This tandem-axle, eight-wheeled draw-bar trailer was designed for use, among others, with the FV12101 Scammell 'Constructor', providing a general service carrier for use with RE equipment, and other items up to the maximum load capacity. A standard draw-bar was provided, and there were winch rope fairleads and a strong point for use with a snatch block, to allow disabled plant, or other dead loads to be winched aboard. Loading was from the rear, by removing the detachable rear axle units; loading ramps were carried on the upper portion of the trailer deck.

Manufacturers
British Trailers & Co Ltd; Hands Ltd; Taskers of Andover (1932) Ltd.

Nomenclature
FV3621. Trailer, 20 ton, 8TW/2LB, low loading.

Dimensions and weight
Length: 9379mm (excluding draw bar).
Width: 2591mm.
Height: 762mm.
Wheelbase: 7480mm
Track: outer, 2172mm; inner, 673mm.
Weight: laden, 31,563kg; unladen, 11,200kg.
Capacity: 20,364kg.

Suspension and axles
Semi-elliptical springs, front and rear; two sets of twin wheels on each axle.

Braking system
Twin air-line pressure braking system designed to be coupled to the tractor; mechanical hand brake, operating on front wheels only.

Road wheels
Four-piece disc wheels, 10.00x20in, mounting standard road-pattern tyres.

FV35003 10 ton trailer for engineering plant
Twin-axle, four-wheeled trailer, designed for carrying excavators and other earth-moving equipment. A standard draw bar was fitted and suitable tractors might have included the FV12101 Scammell 'Constructor'.

The rear axle was easily removable to enable loading from the rear.

Manufacturers
Eagle Engineering Co Ltd.

Nomenclature
FV35003. Trailer, 10 ton, 4TW/2LB, low loader, excavator transporter.

Dimensions and weight
Length: 8832mm.
Width: 2362mm.
Height: 1676mm.
Wheelbase: 5788mm.
Track: 1829mm.
Weight: laden, 15,695kg; unladen, 5534kg.
Capacity: 10,161kg.

Suspension and axles
Semi-elliptical springs, front and rear; twin wheels on each axle.

Braking system
Twin air-line pressure braking system designed to be coupled to the tractor; mechanical hand brake, operating on the front wheels only.

Road wheels
Three-piece disc wheels 9.00x20in, mounting standard road-pattern tyres.

SEMI-TRAILERS

Semi-trailers are designed to be towed by a fifth-wheel tractor; the trailer being attached to the tractor by means of a rotating turntable (the so-called 'fifth wheel'), generally situated immediately over the rear axle bogie. The fifth wheel forms a coupling between the two units, and provides the steering action for the trailer.

These days, the fifth-wheel equipment tends to be both standardised, and easily detachable, which means that one tractor can be used with any number of trailers from a multiplicity of manufacturers, but back then, it was not uncommon for the trailer to be produced by, and designed only for one particular tractor. On top of that, the trailers were also not always easily detachable.

The semi-trailer reduces both the weight and length of the rig, but also reduces the off-road performance, and if the trailer-to-tractor coupling is not carefully designed, the centre of gravity can be high, leading to problems with stability off the highway.

The geometry of the steering action generally dictates that the trailer will cut-in tight on turns, but this is dictated by the position of the fifth-wheel coupling in relation to the rear wheels of the tractor.

Although semi-trailers are almost invariably simply 'trailed' behind the tractor, there was some feasibility work carried out in relation to powered semi-trailers for off-road use in conjunction with the FV1200 Series (see Chapter 4.3).

FV3001/FV3005 60 ton tank-transporter semi-trailer
Both these trailers were designed for use with the Thornycroft 'Antar' FV12002/FV12004 Mk 2 and Mk 3 fifth-wheel tractors; the original model was designated FV3001, the FV3005 was a later development.

With a capacity of 60 tons, both were intended for transporting tanks, AFV's and other dense stores up to the maximum designated loading; later production examples were metricated to 60,000kg capacity but the actual trailer was not modified. Extensive use was made of alloy steel to keep the dead weight to a minimum, and the trailers were designed with a long wheelbase to reduce the axle loading imposed on military bridges by the complete train.

A standard detachable fifth-wheel coupling was used.

Loading was from the rear, by means of hydraulically-operated folding ramps and jacks. Retractable legs and braces at the front of the load platform were used to support the uncoupled trailer. To facilitate loading disabled AFV's, both models were fitted with winch rope guide rollers at front and rear.

Manufacturers
Joseph Sankey & Sons.

Nomenclature
FV3001. Semi-trailer, 60 ton, 8TW/2LB, tank transporter (road).
FV3005. Semi-trailer, 60 ton (later 60 tonne), 8TW/2LB, tank transporter (road).

Dimensions and weight
Length: FV3001, 12,510mm; FV3005, 12,687mm.
Width: 3660mm.
Height: FV3001, 3000mm; FV3005, 2934mm.
Wheelbase: 9449mm (measured to centre of kingpin).
Track: outer, 2920mm; inner, 940mm.
Height of fifth wheel from ground: 1600mm.
Weight: laden, 78,940kg; unladen, 18,940kg.
Capacity: 60,000kg.

Suspension and axles
Walking beam suspension; four pairs of wheels on each axle.

Braking system
Twin air-line pressure-assisted braking system designed to be coupled to the tractor; manually-operated hand brake.

Road wheels
Four-piece disc wheels, 13.00x20in, mounting standard road-pattern tyres. Some examples were fitted with an automatic puncture alarm system.

The trailer was designed to carry one spare wheel for the tractor, and one for the trailer itself, on the forward swan neck.

FV3011 50 ton tank-transporter semi-trailer
Twin-axle, eight-wheeled semi-trailer designed for use with the Thornycroft 'Antar' FV12002/FV12004 Mk 2 and Mk 3 fifth-wheel tractors. Intended for the movement of tanks, AFV's and other dense stores up to 50 tons in weight (later upgraded to 50,000kg). The trailer was constructed from standard structural members to minimise production costs and was fitted with a standard detachable fifth-wheel coupling.

Loading was from the rear, by means of folding ramps and manually-operated hydraulic jacks. Retractable legs and braces at the front of the load platform were used to support the uncoupled trailer. Winch rope guide rollers were provided at front and rear to facilitate loading disabled AFV's.

Manufacturers
Joseph Sankey & Sons; GKN Group Services Ltd; Taskers of Andover (1932) Ltd.

Nomenclature
FV3011. Semi-trailer, 50 ton, 8TW/2LB, tank transporter.

Dimensions and weight
Length: 11,703mm.
Width: 3353mm.
Height: 3048mm.
Wheelbase: 8534mm (measured to centre of kingpin).
Track: outer, 2698mm; inner, 883mm.
Height of fifth wheel from ground: 1600mm.
Weight: laden, 64,240kg; unladen, 14,240kg.
Capacity: 50,000kg.

Suspension and axles
Walking beam suspension; four pairs of wheels on each axle.

Braking system
Twin air-line pressure-assisted hydraulic braking system designed to be coupled to the tractor; manually-operated hydraulic hand brake.

Road wheels
Four-piece disc wheels, 12.00x20in, mounting standard road-pattern tyres. The trailer was designed to carry one spare wheel for the tractor, and one for the trailer itself, on the forward swan neck.

FV3541 30 ton semi-trailer for engineering plant
This tandem-axle, eight-wheeled semi-trailer was designed for use with the FV12102/FV12105 Scammell 'Constructors', providing a vehicle train suitable for transporting heavy engineering plant, bridge members, and pole cable supports up to 12m in length. A standard detachable fifth-wheel coupling was used.

Loading was from the rear, by means of folding ramps and hydraulic jacks designed to be connected to the tractor hydraulic system. On early examples, there were two ramps, running the full-width of the trailer, and centre-hinged to reduce the overall height when in the upright position; on later examples, the ramps were modified to provide three fixed units, again running across the full width of the trailer. Retractable support legs were provided at the front of the load platform for use when the trailer was uncoupled from the tractor.

Manufacturers
Cranes (Dereham) Ltd; Taskers of Andover (1932) Ltd.

Nomenclature
FV3541. Semi-trailer, 30 ton, 8TW/2LB, rear-loader, RE plant.

Dimensions and weight
Length: Taskers, 12,287mm; Cranes, 12,402mm.
Width: Taskers, 3200mm; Cranes, 3048mm.
Height: Taskers, 2743mm; Cranes, 3708mm.
Wheelbase (measured to centre of kingpin): Taskers, 9258mm; Cranes, 9360mm.
Track: outer, 2597mm; inner, 806mm.
Height of fifth wheel from ground: Taskers, 1549mm; Cranes, 1449mm.
Laden weight: Taskers, 46,760kg; Cranes, 45,234kg.
Unladen weight: Taskers, 15,890kg; Cranes, 14,364kg.
Capacity: 30,870kg.

Suspension and axles
Walking beam suspension; four pairs of twin wheels on each axle.

Braking system
Twin air-line pressure braking system designed to be coupled to the tractor; mechanical hand brake.

Road wheels
Three-piece disc wheels, 10.00x15in, mounting standard road-pattern tyres.

DUMMY AXLE and DOLLY TRAILERS

FV3561 10/30 tonne dummy axle trailer
One of the problems of producing massive trucks is that should there be a mechanical failure, a similarly-massive recovery tractor is required to provide assistance. When the 'Antar' was introduced, there was no recovery tractor

with sufficient capacity to provide a suspended tow, and obviously some other approach was called for.

The dummy axle trailer was a simple means of using a draw-bar tractor as a recovery vehicle. The trailer was of single-axle design, with two pairs of wheels; a small supporting dolly axle at the front provided stability for the uncoupled trailer.

There was a draw-bar hitch at the front, with a fixed-length, hydraulic jib mounted on the trailer chassis, and used to lift the casualty. An 'A' frame was provided at the rear to allow coupling to the disabled vehicle. This device enabled one 'Antar' to provide assistance to another on suspended tow.

Manufacturers
Transport Equipment (Thornycroft) Ltd; Royal Ordnance Factory (ROF), Nottingham.

Nomenclature
FV3561. Trailer, 10/30 tonne, dummy axle, recovery.

Dimensions and weight
Length: 3581mm.
Width: 2388mm.
Height: 3302mm.
Wheelbase: 3099mm (measured to centre of kingpin).
Track: 1778mm.
Weight: laden, 13,370kg; unladen, 3250kg.
Capacity: 10,182kg.

Suspension and axles
Unsprung; two pairs of twin wheels on single axle.

Braking system
Twin air-line pressure braking system designed to be coupled to the tractor; mechanical hand brake.

Road wheels
Three-piece disc wheels, 8.00x15in, mounting standard road-pattern tyres.

FV2751 10 ton dolly trailer converter

No matter how carefully the logistics of any operation were planned, there would inevitably be situations where semi-trailers were scheduled for movement, with only ballast tractors available to move them. The dolly converter trailer was a simple means of providing a fifth-wheel coupling for a draw-bar tractor.

The trailer consisted of a single axle, with a draw-bar connection for the tractor, and with a standard fifth wheel mounted to the trailer chassis. The trailer would be coupled to the tractor, and the semi-trailer connected to the fifth-wheel in the normal way. The only disadvantage was that the length and weight of the outfit was extended by some 2m, and the two trailer pivot points must have made reversing very interesting.

Manufacturers
Cranes (Dereham) Ltd.

Nomenclature
FV2751. Dolly trailer converter, 10 ton, 2W/2LB.

Dimensions and weight
Length: 3861mm.
Width: 2540mm.
Height: 1524mm.
Track: 1778mm.
Weight: laden, 10,170kg; unladen, 2190kg.
Capacity: 7980kg.

Suspension and axles
Semi-elliptical springs and shock absorbers; single axle design.

Braking system
Twin air-line pressure braking system designed to be coupled to the tractor.

Road wheels
One-piece disc wheels, 8.00x15in, mounting standard road-pattern tyres.

OTHER TRAILERS

Alongside these heavy equipment trailers, there was also a large number of specialised trailers and semi-trailers intended for other, more-particular transport applications, many of which may have been used with the tractors described in this book. Examples include:

FV2706	Semi-trailer, 10 ton, missile transporter
FV2861	Trailer, 5 ton, heavy bridging
FV3242	Semi-trailer, cargo, 20 ton, commercial pattern, low mobility
FV3501	Trailer, 10 ton, 4TW/2LB, low loader, cargo
FV3542	Semi-trailer, 30 ton, 8TW/2LB, armoured vehicle launched bridge (AVLB)
FV3552	Trailer, 6 ton, 8W/2LB, aircraft transporter
FV3681	Trailer, 50 ton, 16TW/8LB, aircraft salvage trolley
FV3682	Trailer, 50 ton, 4TW, aircraft salvage, nose supporting trolley
FV3751	Trailer, 1W, fuel carrier, for Centurion
FV35001	Trailer, 10 ton, low loading, flat platform, 4TW
FV35002	Trailer, 7.5 ton, 4W/2LB, 1500 gal, water tanker

Many specialised semi-trailers also existed in the 5-10 ton range, purpose-designed for applications such as field offices, radar stations, photographic processing laboratories, print rooms, oxygen and acetylene plant, flight simulators, and parachute drying.

DOCUMENTATION

Technical publications

Report
FVRDE Report FT/B 345. Interim report on semi-trailer 60 ton, 8TW/2LB, tank transporter, FV3001.

Specification
FVRDE Specification 9056. Trailer, 50 ton, tank transporter, 16 twin-wheels, 4 line brake, FV 3601.

User handbooks
FV3001. Semi-trailer, 60 ton, tank transporter, 8 twin-wheels, Sankey, Mk 1. WO code 11218.
FV3011. Semi-trailer, 50 ton, tank transporter, 8 twin-wheels, 2 line brake, Taskers/Sankey. WO code 13290.
FV3541. Semi-trailer, 30 ton, RE plant, rear loading, 8TW, 2LB, Cranes. WO code 13293.
FV3601. Trailer, 50 ton, tank transporter, 16 twin-wheels, 4 line brake, Mks 1, 2 and 3, Cranes/Dyson. Army code 1879.

Servicing schedules
FV3001. Semi-trailer, 60 ton, tank transporter, 8 twin-wheels, Sankey, Mk 1. WO code 10986.
FV3011. Semi-trailer, 50 ton, tank transporter, 8 twin-wheels, 2 line brake, Taskers/Sankey. WO code 13163.
FV3541. Semi-trailer, 30 ton, RE plant, rear loading, 8TW, 2LB, Cranes. WO code 13341.
FV3601. Trailer, 50 ton, tank transporter, 16 twin-wheels, 4 line brake, Cranes/Dyson. WO code 10626.

Parts lists
FV3001. Semi-trailer, 60 ton, tank transporter, 8 twin-wheels, Sankey, Mk 1. WO codes 11785, 13441, 17813.
FV3011. Semi-trailer, 50 ton, tank transporter, 8 twin-wheels, 2 line brake, Taskers/Sankey. WO code 13327.
FV3541. Semi-trailer, 30 ton, RE plant, rear loading, 8TW, 2LB, Cranes. WO code 13223.
FV3601. Trailer, 50 ton, tank transporter, 16 wheeled, No 1. Chilwell catalogue 34/178.
FV3601. Trailer, 50 ton, tank transporter, 16 twin-wheels, 4 line brake, Mks 1, 2 and 3, Cranes/Dyson. WO code 6867.

Technical handbooks
General:
Load-carrying trailers and trailer chassis. Air publication 4675A, volumes 1 and 6.

Data summary:
FV3001. Semi-trailer, 60 ton, tank transporter, 8 twin-wheels, Sankey, Mk 1. EMER U390/2.
FV3011. Semi-trailer, 50 ton, tank transporter, 8 twin-wheels, 2 line brake, Taskers/Sankey. EMER U390/3.
FV3541. Semi-trailer, 30 ton, RE plant, rear loading, 8TW, 2LB, Cranes. EMER U390/4.
FV3601. Trailer, 50 ton, tank transporter, 16 twin-wheels, 4 line brake, Cranes/Dyson. EMER U490/1.

Technical description:
FV3001. Semi-trailer, 60 ton, tank transporter, 8 twin-wheels, Sankey, Mk 1. EMER U392/2.
FV3011. Semi-trailer, 50 ton, tank transporter, 8 twin-wheels, 2 line brake, Taskers/Sankey. EMER U392/3.
FV3541. Semi-trailer, 30 ton, RE plant, rear loading, 8TW, 2LB, Cranes. EMER U392/4, EMER U392/4 supplement 1.
FV3601. Trailer, 50 ton, tank transporter, 16 twin-wheels, 4 line brake, Cranes/Dyson. EMER U492/1.

Repair manuals
Unit repairs:
FV3001. Semi-trailer, 60 ton, tank transporter, 8 twin-wheels, Sankey, Mk 1. EMER U393/2.
FV3011. Semi-trailer, 50 ton, tank transporter, 8 twin-wheels, 2 line brake, Taskers/Sankey. EMER U393/3.
FV3541. Semi-trailer, 30 ton, RE plant, rear loading, 8TW, 2LB, Cranes. EMER U393/4.
FV3601. Trailer, 50 ton, tank transporter, 16 twin-wheels, 4 line brake, Cranes/Dyson. EMER U493/1.

Field and base repairs:
FV3541. Semi-trailer, 30 ton, RE plant, rear loading, 8TW, 2LB, Cranes. EMER U694/4.

Modification instructions
FV3001. Semi-trailer, 60 ton, tank transporter, 8 twin-wheels, Sankey, Mk 1. EMER U397/2, instructions 1-14.
FV3011. Semi-trailer, 50 ton, tank transporter, 8 twin-wheels, 2 line brake, Taskers/Sankey. EMER U397/3, instructions 1-6.

Miscellaneous instructions
FV3011. Semi-trailer, 50 ton, tank transporter, 8 twin-wheels, 2 line brake, Taskers/Sankey. EMER U399/3, instruction 1.
FV3601. Trailer, 50 ton, tank transporter, 16 twin-wheels, 4 line brake, Cranes/Dyson. EMER U499/1, instruction 1.

Complete equipment schedules
FV3001. Semi-trailer, 60 ton, tank transporter, 8 twin-wheels, Sankey, Mk 1. WO code 33861.
FV3011. Semi-trailer, 50 ton, tank transporter, 8 twin-wheels, 2 line brake, Taskers/Sankey. WO code 38864.
FV3541. Semi-trailer, 30 ton, RE plant, rear loading, 8TW, 2LB, Cranes. WO code 33865.

Tools and equipment
FV3001. Semi-trailer, 60 ton, tank transporter, 8 twin-wheels, Sankey, Mk 1. WO code 18497, table 1225.

ALPHABETIC INDEX
Numeric entries in italic indicate photographs

4in heavy anti-aircraft gun 121
5.5in howitzer 21
5.5in howitzer 66
7.5in howitzer 66
8in howitzer 66, 68
980/981 *see Diamond T*

A

ACV 10, 140, 148
AEC 9, 56, 140, 148, *see also specific entries*
    2AV690 engines 23, 26
    2AV760 engines 21, 23, 26
    A223 engines 21, 26
    AV690 engines 21, 23, 26
    AV760 engines 26
    diesel engines 148
    low-profile artillery tractors *16*
    Mammoth Major 10
    Marshal 13, 20
    Matador *see Matador*
    Militant *see Militant*
    Y type 10
Aircraft trailers 161
    FV3552 165
    FV3681 165
    FV3682 165
Albion 7, 31, 56, *see also specific entries*
    EN244 engines 35
    EN248B engines 37, 41
    FV14000 39 *39*
Albion CX22S 33-35, *36*
    axles 35
    bodywork 35
    braking system 35
    chassis 35
    development 33
    electrical equipment 35
    engines 35
    nomenclature 34
    semi-trailers 158
    steering gear 35
    suspension 35
    transmission 35
    wheels 35
    winch 35
Albion CX24S 33, 35-39, *37*, 109
    axles 37
    bodywork 38
    braking system 38
    chassis 38
    development 35
    electrical equipment 39
    engines 37
    fifth-wheel tractors 35, *37*
    nomenclature 36
    problems 36
    semi-trailers 36

    steering gear 38
    suspension 37
    transmission 37
    wheels 38
    winch 39
Albion CX33 33, *37*, 41-42
    axles 42
    bodywork 42
    braking system 42
    chassis 42
    engines 41
    nomenclature 41
    steering gear 42
    suspension 42
    transmission 42
    wheels 42
    winch 42
Albion FT15N/FT15NW 33, *38-39*, 39-41
    axles 40
    bodywork 39
    braking system 40
    chassis 40
    development 39
    electrical equipment 41
    engines 40
    nomenclature 39
    steering gear 39, 40
    suspension 40
    transmission 40
    wheels 40
    winch 41
Alvis Stalwart 143
Antar 7, 8, 48, 61, 94, 96, *100*, 104, 105, 117, 122, 141-154, *144-149*
    axles 151
    ballast tractors 141
    bodywork 152
    bolt-on jib 148
    braking system 151
    chassis 152
    commercial 143
    development 143
    documentation 153
    electrical equipment 152
    engines 149-150
    fifth-wheel tractors 141
    modifications 148
    nomenclature 146
    production 146
    publications 153
    recovery equipment 148
    semi-trailers 158, 163, 164
    steering gear 151
    suspended tow 164
    suspension 151
    trailers 148, 162
    transmission 151
    wheels 152

INDEX

winch 153
Armoured cars, Pioneer *109*, 111
Artillery tractors
  'atomic' howitzer 65
  AEC low-profile *16*
  Albion 33-35, *36*
  Albion FT15N/FT15NW 39
  Constructor 118, 119
  FV1201 93
  Mack 21, 66
  Martian 65, 66
  Matador 11
  Militant 19
  Pioneer 107, 109
Associated Commercial
Vehicles Group 10
Associated Daimler Company 10
Austin Champ 21
Axles
  Albion CX22S 35
  Albion CX24S 37
  Albion CX33 42
  Albion FT15N/FT15NW 40
  Antar 151
  Constructor 125
  Diamond T 52
  Explorer 135
  FV1003 63
  Martian 72
  Matador 16
  Militant 27
  Pioneer 114

B

'B' range vehicles 7, 117, 129
Ballast tractors
  Antar 141
  Constructor 118, 120
  Diamond T 45
  Explorer 130, *134*
  FV1202 93
  Bedford 56, 118
  Bison 56
Bodywork
  Albion CX22S 35
  Albion CX24S 38
  Albion CX33 42
  Albion FT15N/FT15NW 39
  Antar 152
  Constructor 126
  Diamond T 53
  Explorer 137
  FV1003 64
  FV1201 101
  Martian 74
  Matador 17
  Militant 28
  Pioneer 115
Bofors gun 21
Bonallack 146, 152
Braking system
  Albion CX22S 35
  Albion CX24S 38
  Albion CX33 42
  Albion FT15N/FT15NW 40

Antar 151
Constructor 126
Diamond T 52
Explorer 136
FV1003 63
FV1201 100
Martian 73
Matador 16
Militant 27
Pioneer 114
Bridge capacities 6, 59
British Crane and
Excavator Corporation 25
British Mk 1/Mk 2 trailers 158, *158*, 160
British Motor Holdings 56
British Trailers 123, 124, 162
  semi-trailers 149
Brockhouse trailers 26, 161
Burton, C E 143

C

Centurion 57, 112
Champ 21, *61*
Chassis
  Albion CX22S 35
  Albion CX24S 38
  Albion CX33 42
  Albion FT15N/FT15NW 40
  Antar 152
  Constructor 126
  Diamond T 52
  Explorer 136
  FV1003 64
  FV1201 101
  Martian 74
  Matador 17
  Militant 28
  Pioneer 115
Chertsey exhibitions 22, 60, 121
Churchill 56
CL vehicles 20
Cleaver, Charles 12
Coles 25
Comet 56
Commander 104, 122, 140, 143
Conqueror 56, 57, 97, 141
Constructor *7*, 8, 51, 96, 104, 117-128, *120-124*
  axles 125
  bodywork 126
  braking system 126
  chassis 126
  commercial 119, *120*
  development 118
  documentation 127
  electrical equipment 127
  fifth-wheel tractors 117, *122*, *123*
  modifications 120, 123
  nomenclature 122
  production 121
  publications 127
  recovery tractors 123
  road-surfacing machine 123

semi-trailers 158
steering gear 126
suspension 125
trailers 123, 161, 162
transmission 125
wheels 126
winch 127
Covenantor 56
Crane Fruehauf *see* Cranes
Cranes
  dolly trailers 165
  semi-trailers 149, 164
  trailers 45, 50, 124, 148, 160, 162
Cromwell 56
Crossley 10, 21, 26, 56
  trailers 71, 161
Crusader 56, 104
CT vehicles 5, 8, 20, 57, 129, 141
CX22S *see* Albion CX22S
CX24S *see* Albion CX24S
CX33 *see* Albion CX33

D

DAF 10, 32, 56, 104
Daimler 10, 56
Darlington winch 153
Dennis 56, 93, 94, 99
  Octolat 41
Development
  Albion CX22S 33
  Albion CX24S 35
  Albion FT15N/FT15NW 39
  Antar 143
  'B' range vehicles 7
  Constructor 118
  CT vehicles 8
  Diamond T 47
  FV1003 59
  FV1201 94
  FVRDE 8
  Martian 66
  Matador 12
  Militant 20
  Pioneer 107
Diamond Reo 44
Diamond T *6*, 7, 43, 45-54, *48-50*, 57, 93, 109, 140, 143, 144
  axles 52
  bodywork 53
  braking system 52
  chassis 52
  development 47
  documentation 54
  electrical equipment 53
  engines 51
  fifth-wheel tractors 47, *48*, *49*
  modifications 47
  nomenclature 50
  problems 47
  production 49
  publications 54
  semi-trailers 158
  shortages 41
  steering gear 52

suspension 52
tank tractors 45
tank transporters 45
trailers 50, 159, 160
transmmission 51
wheels 52
winch 53
Diesel engines 7
Documentation
  Antar 153
  Constructor 127
  Diamond T 54
  Explorer 138
  FV1003 64
  FV1201 102
  Martian 75
  Matador 18
  Militant 29
  Pioneer 116
  semi-trailers 166
  trailers 166
Dolly recovery 148
Dolly trailers
  Cranes 165
  FV2751 158, *161*, 165
Double heading *6*, 47, 57
Dragon Wagon 45, 57
Dummy axle trailers 164
  FV3561 158, *161*
Dyson trailers 45, 50, 124, 148, 160, 162

E

Eagle Engineering trailers 163
Edmonds, W A 24
EKA recovery vehicles 20
Electrical equipment
  Albion CX22S 35
  Albion CX24S 39
  Albion FT15N/FT15NW 41
  Antar 152
  Constructor 127
  Diamond T 53
  Explorer 137
  FV1003 64
  FV1201 102
  Martian 75
  Matador 17
  Militant 29
  Pioneer 115
Electrical screening 6
Engineering plant
  semi-trailers 164
  trailers 162
Engines
  AEC 2AV690 23, 26
  AEC 2AV760 21, 23, 26
  AEC A173 14
  AEC A187 14
  AEC A223 21, 26
  AEC AV690 21, 23, 26
  AEC AV760 26
  AEC diesel 148
  Albion CX22S 35
  Albion CX24S 37

INDEX

Albion CX33 41
Albion EN244 35
Albion EN248B 37
Albion EN248E 41
Albion FT15N/FT15NW 40
Antar 149-150
Diamond T 51
diesel 7
FV1003 62
Gardner 6LW 108, 113, 130
Hall-Scott 44
Hercules 44
Hercules DXFE 47, 51
Hercules DXFE supercharged 48, *50*
Leyland 0600 70
Martian 65, 67, 70, 71
Matador 14
Meadows 117, 119
Meteorite 59, 62, 95, 98, 140, 143, 149
Militant 21, 26
multi-fuel 21, 23
multi-fuel problems 24
petrol 7
Pioneer 108, 113
Rolls-Royce 'B' series 65, 129
Rolls-Royce B80 39, 67, 108, 140
Rolls-Royce B81 67
Rolls-Royce C6 44, 48, 51, 118, 119, 124
Rolls-Royce C8 145, 150
Rover Meteorite 59, 62, 95, 98, 140, 143, 149
Scammell-Meadows 6DC-630 124
Scammell-Meadows 6PC-630 117, 119, 124
Scammell-Meadows 6PC-653 129, 133
Explorer 7, 13, 19, 65, 104, 108, 119, 129-138, *132-136*
  axles 135
  ballast tractors 130, *134*
  bodywork 137
  braking system 136
  chassis 136
  commercial 130
  documentation 138
  electrical equipment 137
  fifth-wheel tractors 131, *136*
  nomenclature 132
  production 132
  publications 138
  recovery equipment 131, *136*, 137
  steering gear 135
  suspension 135
  towing tractors 130, *134*
  trailers 132, 160
  transmission 134
  wheels 136
  winch 137

F

Fifth-wheel tractors
  Albion CX24S 35, *37*
  Antar 141
  Constructor 117, *122, 123*
  Diamond T 47, *48, 49*
  Explorer 131, *136*
  FV1003 57
  FV1200 93
  Martian 65
  Pioneer 105, *109, 110*
  trailers *see* semi-trailers
Fluidrive transmission 99
Four Wheel Drive Motor Company 11, 12
Fruehauf trailers 45
FT15N/FT15NW *see* Albion FT15N/FT15NW
Fulton, Norman 32
FV1000 56, 57-64, *60-63*, 93, 94, 117, 141, 144, 148
  semi-trailers 158
FV1001 59
FV1003 56-64, *60-63*, 63
  axles 63
  bodywork 64
  braking system 63
  chassis 64
  development 59
  documentation 64
  electrical equipment 64
  engines 62
  fifth-wheel tractors 57
  nomenclature 61
  production 59
  publications 64
  radio equipment 64
  semi-trailers 61, 64
  steering gear 63
  suspension 63
  transmission 62
  wheels 64
  winch 64
FV1004 59
FV1100 *see Martian*
FV11001 *see Militant*
FV11002 *see Militant*
FV11005 *see Militant*
FV11007 *see Militant*
FV11008 *see Militant*
FV11009 *see Militant*
FV1101 *see Martian*
FV1103 *see Martian*
FV11041 *see Militant*
FV11044 *see Militant*
FV11047 *see Militant*
FV11061 *see Militant*
FV1119 *see Martian*
FV1121 *see Martian*
FV1122 *see Martian*
FV11300 *see Explorer*
FV11301 *see Explorer*
FV1200 56, *62*, 65, 93-102, *96-101*, 117, 140, 141, 144, 148
  fifth-wheel tractors 93

FV12000 *see Antar*
FV12001 *see Antar*
FV12002 *see Antar*
FV12003 *see Antar*
FV12004 *see Antar*
FV12005 *see Antar*
FV12006 *see Antar*
FV12007 *see Antar*
FV1201 93-102, *96-101*
  bodywork 101
  braking system 100
  chassis 101
  development 94
  documentation 102
  electrical equipment 102
  nomenclature 98
  problems 95
  publications 102
  radio equipment 102
  steering gear 100
  suspension 100
  transmission 99
  wheels 100
  winch 102
FV1202 93
FV1203 93
FV1204 93
FV1205 93
FV1206 93
FV1207 93
FV12100 *see Constructor*
FV12101 *see Constructor*
FV12102 *see Constructor*
FV12103 *see Constructor*
FV12104 *see Constructor*
FV12105 *see Constructor*
FV1300 39
FV14000 39
FV14000 *see Albion*
FV1800 *see Champ*
FV200 56, 57
FV214 56
FV2451 148
FV2706 165
FV2751 158, *161*, 165
FV2861 165
FV3001 141, 149, 158, *160*, 163
FV3005 141, 149, 158, 163
FV3011 149, 158, *160*, 164
FV3221 26, 71, 133, 158, *158*, 160
FV3242 165
FV3301 59
FV35001 165
FV35002 165
FV35003 158, *160*, 162
FV3501 165
FV3521 149
FV3541 121, 123, 158, *161*, 164
FV3542 165
FV3551 124, 158, *159*, 161
FV3552 165
FV3561 158, *161*, 164
FV3561 164
FV3601 51, 124, 141, 144, 148, 158, *159*, 162
FV3621 112, 121-124, 158, *159*, 162
FV3681 165
FV3682 165
FV3751 112, 165
FVRDE exhibitions 22, 60, 121

G

Gar Wood winch 53
Gardner 6LW engines 108, 113, 130
General Motors 56
GKN semi-trailers 149, 164
GS vehicles 8, 20, 129, 144
Gun tractors
  Albion 33-35, *36*
  Albion FT15N/FT15NW 39
  Constructor 118
  FV1201 93
  Mack 21, 66
  Martian 65, 66
  Matador 11
  Militant 19
  Pioneer 107, 109
Guy 56

H

Hall-Scott engines 44
Hands (Letchworth)
  semi-trailers 149
  trailers 123, 124, 162
Hardwick's yard 61, 98
Hardy Motors 11, 12
Hercules engines 44
  DXFE engines 47, 51
  DXFE engines supercharged 48, *50*
Highwayman 119
Hippo 56, 65
Hugh, P G 107
Hydraulic Couplings transmission 99

I

Iraq Petroleum Company 143

K

Kightley 95
Kirkstall Forge Engineering 151

L

Lancashire Steam Motor Company 56
Leyland 48, 55, 140, *see also specific entries*
  0600 engines 70
  DAF 10, 32, 56, 104
  Glasgow 32
  Martian *see Martian*
Liberty trucks 44
Light recovery trailers *see FV3221*

INDEX

M
M3 36
M19 45
M20 see Diamond T
M25 45, 57
M26 45
M54 44
M62 44
M9 45, 50, 159
Mack 66
   artillery tractors 21
   gun tractors 21
Mann-Egerton 69
Marsh, C 24
Martian *6*, 13, 20, 21, 56, 65-76, *68-72*, 94, 107
   Australian trials 69
   axles 72
   bodywork 74
   braking system 73
   chassis 74
   commercial 66, 70
   development 66
   documentation 75
   electrical equipment 75
   engines 65, 67, 70, 71
   fifth-wheel tractors 65
   nomenclature 70
   problems 13, 66, 67
   production 69
   publications 75
   recovery equipment 70, *71*
   steering gear 73
   suspension 72
   trailers 70, 160
   transmission 72
   wheels 73
   winch 75
Martlew, C K 24
Matador 7, 10-18, *14-16*, 20, 65, 68, 107
   axles 16
   bodywork 17
   braking system 16
   chassis 17
   development 12
   documentation 18
   electrical equipment 17
   engines 14
   nomenclature 14
   production 11, 13
   publications 18
   steering gear 16
   suspension 16
   transmission 15
   wheels 17
   winch 17
Matilda 56
Maudslay 10, 23, 56
Meadows engines 117, 119
Meteorite engines 59, 62, 95, 98, 140, 143, 149
MEXE 123
Mighty Antar see Antar

Militant 8, 10, 11, 19-30 *22-25*, 130, 132
   axles 27
   bodywork 28
   braking system 27
   chassis 28
   development 20
   documentation 29
   electrical equipment 29
   engines 21, 26
   nomenclature 25
   production 21
   publications 29
   reactor 29
   recovery equipment 28
   recovery vehicles 19
   steering gear 27
   suspension 27
   trailers 26
   transmission 26
   wheels 28
   winch 29
Modifications
   Antar 148
   Constructor 120, 123
   Diamond T 47
   Pioneer 111
Motor Panels 152
Mountaineer 122
Mudlark *61*
Multi-fuel engines 21, 23
   problems 24
Murray, Blackwood T 32

N
Nomenclature
   Albion CX22S 34
   Albion CX24S 36
   Albion CX33 41
   Albion FT15N/FT15NW 39
   Antar 146
   Constructor 122
   Diamond T 50
   Explorer 132
   FV1003 61
   FV1201 98
   Martian 70
   Matador 14
   Militant 25
   Pioneer 110
North, Oliver 107

O
O853 Matador 11
O854 Matador 11
O859 Militant 20
O860 Militant 20

P
Pacific TR1 modified 57, *60*
Park Royal 22, 69
Petrol engines 7
Pioneer 7, 33, 45, 57, 65, 93, 104-116, *108-113*, 117, 119, 122, 129, 140, 143

armoured car *109*, 111
axles 114
bodywork 115
braking system 114
chassis 115
development 107
documentation 116
electrical equipment 115
modifications 111
nomenclature 110
production 109, 110
publications 116
semi-trailers 112, 114, 115, 158
suspension 114
trailers 160
transmission 113
wheels 115
winch 115
Production
   Antar 146
   Constructor 121
   Diamond T 49
   Explorer 132
   FV1003 59
   Martian 69
   Matador 13
   Militant 21
   Pioneer 109, 110
Publications
   Antar 153
   Constructor 127
   Diamond T 54
   Explorer 138
   FV1003 64
   FV1201 102
   Martian 75
   Matador 18
   Militant 29
   Pioneer 116
   semi-trailers 166
   trailers 166
Punctures, trailers 157

R
R100 see Pioneer
Radio equipment
   FV1003 64
   FV1201 102
Radio interference 6
RAF-type trucks 56
RE trailers 162
Reactor, Militant 29
Recovery equipment
   Antar 148
   Constructor 123
   dolly trailers 148
   EKA 20
   Explorer 131, *136*, 137
   Martian 70, *71*
   Militant 28
Recovery tractors
   EKA 20
   Explorer 65, 129
   FV1205 93

Martian 65, 68
Militant 19, 130, 132
Pioneer 107, 109
Red Maid anti-aircraft gun 95
REME 48, 68, 130
Reo 44
Ricardo 10
Road-surfacing machine, Constructor 123
Rogers M9 trailers 14, 45, 159
Rolls-Royce
   'B' series engines 65, 129
   B80 engines 39, 67, 108, 140
   B81 engines 67
   C6 engines 44, 48, 51, 118, 119, 124
   C8 engines 145, 150
Rover Meteorite engines 59, 62, 95, 98, 140, 143, 149
Royal Ordnance Factories 70
   trailers 165
Rubery Owen trailers 26, 71, 161

S
Sankey semi-trailers 149, 163, 164
Scammell 25, 32, 56, 65, 103, 140, *see also specific entries*
   Commander 122, 140, 143
   Constructor see Constructor
   Explorer see Explorer
   Highwayman 119
   Mechanical Horse 118
   Military '14' 110
   Mountaineer 122
   Pioneer see Pioneer
   Super Constructor 122
Scammell winch
   Albion CX22S 35
   Albion CX24S 39
   Albion CX33 42
   Constructor 127
   Explorer 137
   Militant 29
   Pioneer 115
Scammell-Meadows
   6DC-630 engines 124
   6PC-630 engines 117, 119, 124
   6PC-653 engines 129, 133
Scott, W A 96
Semi-trailers 155, 163
   Albion CX22S 158
   Albion CX24S 36
   Antar 158, 163, 164
   British Trailers 149
   Constructor 158
   Cranes 149, 164
   Diamond T 158
   documentation 166
   engineering plant 164
   FV1000 158
   FV1003 61, 64
   FV3001 141, 149
   FV3001 158, *160*, 163
   FV3005 141, 149, 158, 163

INDEX

FV3011 149, 158, *160*, 164
FV3301 59, 61, 64
FV3521 149
FV3541 121, 123, 158, *161*, 164
GKN 149, 164
Hands (Letchworth) 149
Pioneer 105, 112, 114, 115, 158
publications 166
Sankey 149, 163, 164
Shelvoke & Drury 51-53, 112
tank transporters 163-164
Taskers 149, 164
TRCU/20 108
TRCU/30 108
Sewell, Rex 21
Shelvoke & Drury semi-trailers 45, *49*, 51-53, 112
Stalwart 143
Standardisation, vehicles 7
Steering gear
 Albion CX22S 35
 Albion CX24S 38
 Albion CX33 42
 Albion FT15N/FT15NW 39, 40
 Antar 151
 Constructor 126
 Diamond T 52
 Explorer 135
 FV1003 63
 FV1201 100
 Martian 73
 Matador 16
 Militant 27
Stuart M3 tank 36
Super Constructor 122
Suspension
 Albion CX22S 35
 Albion CX24S 37
 Albion CX33 42
 Albion FT15N/FT15NW 40
 Antar 151
 Constructor 125
 Diamond T 52
 Explorer 135
 FV1003 63
 FV1201 100
 Martian 72
 Matador 16
 Militant 27
 Pioneer 114
SV/1S *see Pioneer*
SV/1T *see Pioneer*
SV/2S *see Pioneer*

T

Tank recovery vehicles 41, 45, 56
Tank tractors
 Albion CX33 41
 Diamond T 45
 FV1201 93, 97, *100*
Tank transporters 56
 Albion 33, 35-39, *37*, 41-42
 Albion CX33 41
 Antar 141
 Diamond T 45

FV1003 57
Highwayman 119
Pioneer 105
trailers 159, 162-164
Taskers
 semi-trailers 149, 164
 trailers 123, 124, 161, 162
Tattersall 95
Thornycroft, John Isaac 140
Thornycroft 23, 56, 139, *see also specific entries*
 Antar *see Antar*
 Nubian 140
 trailers 165
 Type J truck 140
Thorpe 95
Tilt, Charles Arthur 44
Towing tractors, Explorer 130, *134*
Townsin, Alan 143
Trailers 155, 158
 aircraft 161
 aircraft FV3552 165
 aircraft FV3681 165
 aircraft FV3682 165
 Antar 148, 162
 British Mk 1/Mk 2 158, *158*, 160
 British Trailers 123, 124, 162
 Brockhouse 161
 Centurion fuel carrier 112
 Constructor 123, 161, 162
 Cranes 124, 148, 160, 162
 Crossley 71, 161
 Diamond T 50, 159, 160
 documentation 166
 dummy axle FV3561 164
 Dyson 124, 148, 160, 162
 Eagle Engineering 163
 engineering plant 162
 Explorer 132, 160
 FV2451 148
 FV2861 165
 FV3221 26, 71, 133, 158, *158*, 160
 FV35001 165
 FV35002 165
 FV35003 158, *160*, 162
 FV3501 165
 FV3551 124, 158, *159*, 161
 FV3552 165
 FV3601 51, 124, 141, 144, 148, 158, *159*, 162
 FV3621 112, 121, 123, 124, 158, *159*, 162
 FV3681 165
 FV3682 165
 FV3751 165
 Hands (Letchworth) 123, 124, 162
 M9 159
 Martian 70, 160
 Militant 26
 Pioneer 160
 publications 166

punctures 157
RE 162
Rogers M9 159
Royal Ordnance Factory 165
Rubery Owen 71, 161
tank transporters 159, 162-164
Taskers 123, 124, 161, 162
Thornycroft 165
Winter-Weiss 159
Transmission
 Albion CX22S 35
 Albion CX24S 37
 Albion CX33 42
 Albion FT15N/FT15NW 40
 Antar 151
 Constructor 125
 Diamond T 51
 Explorer 134
 Fluidrive 99
 FV1003 62
 FV1201 99
 Hydraulic Couplings 99
 Martian 72
 Matador 15
 Militant 26
 Pioneer 113
Transport Equipment (Thornycroft) 140
TRCU/20 *see Pioneer*
TRCU/30 *see Pioneer*
TRMU/20 *see Pioneer*
TRMU/30 *see Pioneer*
Turner winch
 Albion FT15N/FT15NW 41
 Antar 153
 Matador 17
 Militant 29

V

Valentine 13
Vauxhall 39
Vickers 95
 medium tank 108

W

Wading 6
War Office Policy Statement '71' 21
Warner electric braking system
 Matador 17
 Militant 27
Weight
 of AFV's 6
 vehicle, gross 59
Wells 95
Wheels
 Albion CX22S 35
 Albion CX24S 38
 Albion CX33 42
 Albion FT15N/FT15NW 40
 Antar 152
 Constructor 126
 Diamond T 52
 Explorer 136
 FV1003 64

FV1201 100
Martian 73
Matador 17
Militant 28
Pioneer 115
Whitehead, W G 95
Wilde winch 75
Wimpey, Geo 141
Winch
 Albion CX22S 35
 Albion CX24S 39
 Albion CX33 42
 Albion FT15N/FT15NW 41
 Antar 153
 Constructor 127
 Darlington 153
 Diamond T 53
 Explorer 137
 FV1003 64
 FV1201 102
 Martian 75
 Matador 17
 Militant 29
 Pioneer 115
 Scammell 35, 39, 42, 115, 127, 137
 Turner 41, 153
 Wilde 75
Winter-Weiss, trailers 50, 159
WOPS '71' 21
WW2 vehicles 5

INDEX